JACARANDA

GEOGRAPHY ALIVE 8

AUSTRALIAN CURRICULUM | SECOND EDITION

JACARANDA
GEOGRAPHY ALIVE 8

AUSTRALIAN CURRICULUM | SECOND EDITION

COORDINATING AUTHOR

JUDY MRAZ

CONTRIBUTING AUTHORS

JUDY MRAZ

CATHY BEDSON

TERRY McMEEKIN

KATHRYN GIBSON

CLEO WESTHORPE

ANNE DEMPSTER

ALISTAIR PURSER

ALEX ROSSIMEL

DENISE MILES

jacaranda
A Wiley Brand

Second edition published 2018 by
John Wiley & Sons Australia, Ltd
42 McDougall Street, Milton, Qld 4064

First edition published 2013

Typeset in 11/14 pts TimesLTStd

© John Wiley & Sons Australia, Ltd 2018

The moral rights of the authors have been asserted.

National Library of Australia
Cataloguing-in-publication data

Author:	Mraz, Judy, author.
Title:	Jacaranda Geography Alive 8 for the Australian Curriculum / Judy Mraz, Cathy Bedson, Terry McMeekin, Kathryn Gibson, Cleo Westhorpe, Anne Dempster, Alistair Purser, Alex Rossimel, Denise Miles
Edition:	Second edition.
ISBN:	9780730347903 (paperback)
Notes:	Includes index.
Target Audience:	For secondary school age.
Subjects:	Geography — Textbooks. Geography — Study and teaching (Secondary) — Australia. Education — Curricula — Australia.
Other Creators/ Contributors:	Bedson, Cathy, author. McMeekin, Terry, author. Gibson, Kathryn, author. Westhorpe, Cleo, author. Dempster, Anne, author. Purser, Alistair, author. Rossimel, Alex, author. Miles, Denise, author.

Trademarks
Jacaranda, the JacPLUS logo, the learnON, assessON and studyON logos, Wiley and the Wiley logo, and any related trade dress are trademarks or registered trademarks of John Wiley & Sons Inc. and/or its affiliates in the United States, Australia and in other countries, and may not be used without written permission. All other trademarks are the property of their respective owners.

Front cover image: Eric Middelkoop / Shutterstock

Cartography by Spatial Vision, Melbourne and MAPgraphics Pty Ltd, Brisbane

Illustrated by various artists and the Wiley Art Studio.

Typeset in India by diacriTech

This textbook contains images of Indigenous people who are, or may be, deceased. The publisher appreciates that this inclusion may distress some Indigenous communities. These images have been included so that the young multicultural audience for this book can better appreciate specific aspects of Indigenous history and experience.

It is recommended that teachers should first preview resources on Indigenous topics in relation to their suitability for the class level or situation. It is also suggested that Indigenous parents or community members be invited to help assess the resources to be shown to Indigenous children. At all times the guidelines laid down by the relevant educational authorities should be followed.

Printed in Singapore
M22R-083_221221

CONTENTS

7 Fieldwork inquiry: How does a waterway change from source to sea? 162

UNIT 2 CHANGING NATIONS 165

8 Urbanisation 167

9 The rise and rise of urban settlements 198

10 Planning Australia's urban future   234

11 Geographical inquiry: Investigating an Asian megacity   254

HOW TO USE

the *Jacaranda Geography Alive* resource suite

For more effective learning, the *Jacaranda Geography Alive* series is now available on the learnON platform. The features described here show how you can use *Jacaranda Geography Alive* to optimise your learning experience.

'Geographical concepts' is a valuable reference section.

'Each concept is clearly defined.'

A series of activities to build and develop your understanding of each concept is provided.

A variety of useful resources support the explanations.

Activities provide you with an opportunity to apply all of the seven concepts.

Linking to *myWorld Atlas* will deepen your understanding.

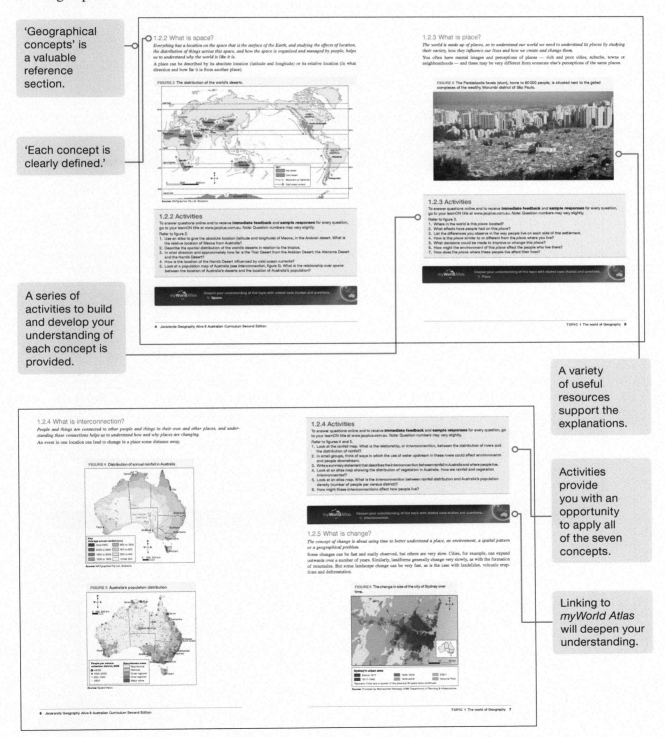

A thought-provoking topic opener sets the scene for your inquiry.

Questions raise issues, link the unit to your life, and prompt you to think about what you already know and feel about the unit.

Evocative and informative images stimulate interest and discussion.

A sequence for your inquiry.

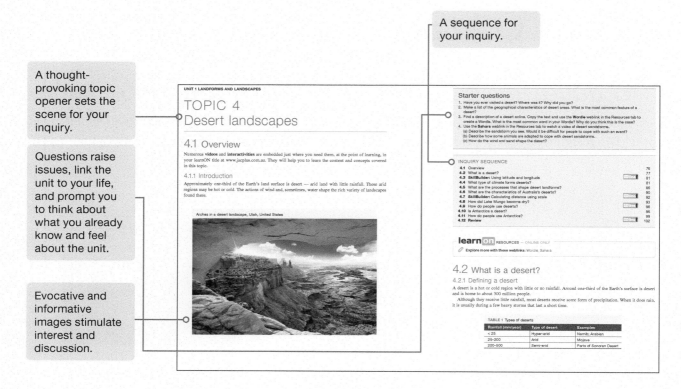

Easily identifiable visual material is referenced in the text and in activities.

Each section begins with a clearly identifiable subtopic number and inquiry question.

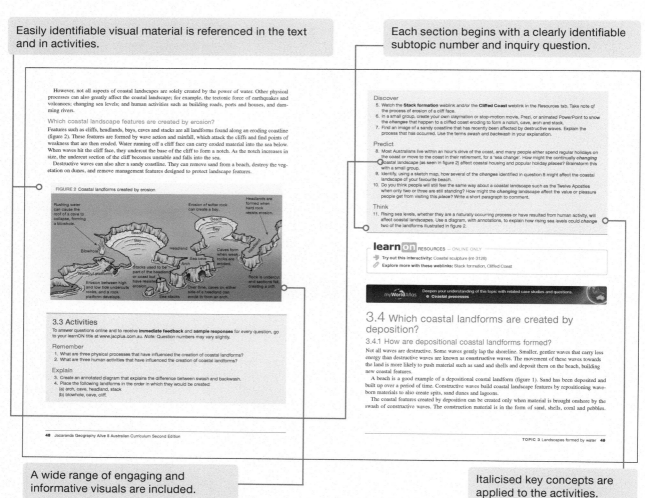

A wide range of engaging and informative visuals are included.

Italicised key concepts are applied to the activities.

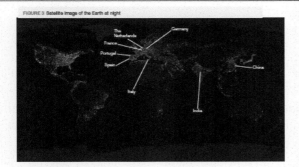
FIGURE 3 Satellite image of the Earth at night

The Netherlands
Germany
France
Portugal
Spain
China
Italy
India

9.2 Activities

To answer questions online and to receive **immediate feedback** and **sample responses** for every question, go to your learnON title at www.jacplus.com.au. Note: Question numbers may vary slightly.

Remember

1. How is a city different from a town or a village?
2. What do the bright lights in figure 1 show?

Explain

3. Study figures 1 and 3 and refer to a political map in your atlas. Which of the following statements are true and which are false? Rewrite the false statements to make them true.
   (a) Japan is a highly populated country with many cities.
   (b) The west coast of the United States is more densely populated than the east coast.
   (c) The Amazon rainforest does not have any settlements.
   (d) The eastern region of China has more cities than the western region.
   (e) The main city settlements in Australia are along the east coast.
   (f) The distribution of cities across Europe is uneven.
4. After completing the 'Describing photographs' SkillBuilder in subtopic 9.3, complete the following questions about figure 2.
   (a) Describe the foreground and background shown in the photograph.
   (b) List the natural and human characteristics shown in the photograph.
   (c) What does this photograph show about urban environments? How has the urban environment changed the natural environment?
   (d) How might the changes described in part (b) lead to an increased risk of erosion? (See topic 2 for information on erosion processes.)
   (e) Imagine that the population of this city continues to increase. Describe what might happen to the land in the future.
   (f) Do you think that all land surrounding cities should be able to be taken up by buildings? Why or why not?

(g) Investigate the place where you live. Are there land-use zones that cannot be built upon, such as 'green wedges'? Where are they and why are they there? Do you think they should be protected from development? Justify your answer.

Discover

5. Use a political map in your atlas and figure 1 to identify the following.
   (a) The Nile River
   (b) The Trans-Siberian railway from Moscow to Vladivostok
   (c) Highways linking cities in the western and eastern United States
   (d) The Himalayan mountain range
6. Go to the **World City Populations** weblink in the Resources tab. Work as a team of five and investigate the change in city population in different continents. Discuss which regions each member will investigate and record maps, data and graphs. Report your findings back to the group.

Predict

7. 'As the world's population continues to increase, cities will spread into the darker regions shown in figure 1.' State whether you agree or disagree with this statement, providing reasons for your decision.

learnon RESOURCES — ONLINE ONLY

Explore more with this weblink: World City Populations

9.3 SkillBuilder: Describing photographs

WHAT IS MEANT BY 'DESCRIBING A PHOTOGRAPH'?

A description is a brief comment (up to a paragraph) on a photograph, identifying and communicating features from a geographic point of view. As geographers, we use our understanding of the world to interpret the image and tell others about the main features or information the photograph reveals.

FIGURE 1 A modern city environment.

Go online to access:
- a clear step-by-step explanation to help you master the skill
- a model of what you are aiming for
- a checklist of key aspects of the skill
- a series of questions to help you apply the skill and to check your understanding.

learnon RESOURCES — ONLINE ONLY

Watch this eLesson: Watch this video to learn how to describe photographs. Searchlight ID: eles-1660

Try out this interactivity: Use this interactivity to learn how to describe photographs. Searchlight ID: int-3156

> **References to online material are provided.**

> **SkillBuilders develop and model key geographical skills in context.**

> **The Fieldwork inquiry and Geographical inquiry provide you with an opportunity to develop your inquiry skills in the field and through research.**

UNIT 2 CHANGING NATIONS

TOPIC 11
Geographical inquiry: Investigating an Asian megacity

11.1 Overview

Numerous **videos** and **interactivities** are embedded just where you need them, at the point of learning, in your learnON title at www.jacplus.com.au. They will help you to learn the content and concepts covered in this topic.

11.1.1 Scenario and your task

The latest liveability report for Asian megacities has been released, and residents are concerned. Populations are increasing by between one and five per cent every year, putting city infrastructure under extreme pressure.

City authorities have commissioned your team to put together a website increasing awareness of the characteristics of the Asian megacity and informing residents of current and newly proposed sustainable development planning initiatives.

Your task

Your team has been put in charge of creating a website designed to inform the residents of an Asian megacity about its characteristics. Each city will be different depending on its location, wealth or poverty, size and climate. Your investigations need to ensure that the audience can gain a comprehensive understanding of both population characteristics and city characteristics, and that any urban problems are presented. A key feature of your website will be to cover any urban solutions and innovations that are currently being implemented in your megacity.

11.2 Process

11.2.1 Process

- You can complete this project individually or invite members of your class to form a group.
- **Planning:** You will need to research the characteristics of your chosen Asian megacity. Research topics that have been loaded in the Resources tab to provide a framework for your research include: location and city characteristics (main economy, tourism, culture); population characteristics (migrants and migration, languages, religion); and urban problems, solutions and innovations. Choose a number of these topics to include in your website and ensure you add your own. Divide the research tasks among the members of your group.

11.2.2 Collecting and recording data

Begin by discussing with your group what you might already know about your chosen Asian megacity. Then discuss the information you will be looking for and where you might find it. To discover extra information about life in your Asian megacity, find at least three sources other than the textbook. At least one of these should be an offline source such as a book or an encyclopaedia. Remember that you will need to choose specific keywords to enter into your search engine to find other data. You can view and comment on other group members' articles and rate the information they have entered.

11.2.3 Analysing your information and data

- You now need to decide what information to include in your website. Maps to show location, graphs, tables and lists to illustrate data, and images and photos with annotations (descriptive notes) should all be included. Each of these should also have a written description. You should make sure that you have addressed each of the following points.
  1. Describe the pattern of distribution on each of the maps or satellite images you have drawn or collected.
  2. What are the main characteristics of your city?
  3. How has your city changed over time? Is information available on how it is predicted to change in the future?
  4. For what reasons are people attracted to move to this city?
  5. What are the main problems in this city? Are there any solutions being introduced to try to overcome these problems?
- Download the website model and website-planning template to help you build your website.
- Use the website-planning template to create design specifications for your site. You should have a home page and at least three link pages per topic. You might want to insert features such as 'Amazing facts' and 'Did you know?' into your interactive website. Remember the three-click rule in web design — you should be able to get anywhere in a website (including back to the homepage) with a maximum of three clicks.

Inside your *Jacaranda Geography Alive learnON*

Jacaranda Geography Alive learnON is an immersive digital learning platform that enables real-time learning through peer-to-peer connections, complete visibility and immediate feedback. It includes:

- a wide variety of embedded videos and interactivities to engage the learner and bring ideas to life
- the **Capabilities** of the Australian Curriculum, available in and throughout the course in activities and **Discussion** widgets
- links to the *myWorld Atlas* for media-rich case studies
- sample responses and immediate feedback for every question
- **SkillBuilders** that present a step-by-step approach to each skill, where each skill is defined and its importance clearly explained
- collaborative activities
- and much more.

ACKNOWLEDGEMENTS

The authors and publisher would like to thank the following copyright holders, organisations and individuals for their assistance and for permission to reproduce copyright material in this book.

Images

• Alamy Australia Pty Ltd: **5**/Pulsar Images; **9** (bottom), **91** (middle)/National Geographic Creative; **12** (top left), **236** (top)/ Michael Willis; **13** (bottom), **56**/Patrick Strattner / fStop Images GmbH; **21** (bottom)/© Mark A. Johnson; **24** (top)/ Bill Bachman; **29**/Global Warming Images; **38**/© Martin Norris Travel Photography; **51**/© robertharding; **53**/David Wall; **55**/robertharding; **60** (top)/keith morris; **68** (top right)/Bill Bachman; **91** (top left)/© Ingo Oeland; **97**/Accent Alaska.com; **108**/redbrickstock.com; **123**/© DOD / S.Dupuis; **137**/Amazon-Images; **153** (bottom)/© Friedrich Stark; **155** (top)/DEEPU SG; **157** (bottom)/Sue Cunningham Photographic; **175**/© Andrew Sole; **205** (middle right)/© Mark Thomas; **220; 230, 249**/ Chris Willson; **241** (bottom)/Australia; **250**/© TGB • Alamy Stock Photo: **246** (top right)/Alamy / Global Warming Images / Ashley Cooper • Alex Rossimel: **60** • Aussie Kanck: **231** (top), **232**/Photo © A Kanck, Quality Freelancing • Australian Antarctic Division: **101**/IA39475 Aerial view of Davis station, Vestfold Hills. Photograph © Darryl Seidel, courtesy Australian Antarctic Division • Australian Bureau of Statistic: **172** (bottom right)/Based on Australian Bureau of Statistics data from 3238.0.55.001 - Experimental Estimates of Aboriginal and Torres Strait Islander Australians, Jun 2006; **189** (bottom); **248**/ http://www.abs.gov.au/ausstats/abs@.nsf/Lookup/by%20Subject/1301.0~2012~Main%20Features~Households%20and%20 families~56; **248**/ http://www.abs.gov.au/ausstats/abs@.nsf/Lookup/by%20Subject/1301.0~2012~Main%20Features~Types%20 of%20Dwellings~127 • Australian Wildlife Conservanc: **41**/Wayne Lawler • Coastal Studies Institute, LSU: **70**/Mike Blum & Harry H. Roberts / Coastal Studies Institute LSU Baton Rouge, LA • Creative Commons: **10** (bottom), **226** (bottom)/ Bahnmoeller; **181** (top)/Graph based on data from the ABS and US Population Bureau; **189** (toop)/© Commonwealth of Australia; **190** (middle)/Department of Immigration and Border Protection; **243**/Material/information courtesy of Department of Climate Change and Energy Efficiency • Food and Agriculture Organization of the United Nations: **245** (top)/Source: Food and Agriculture Organization of the United Nations, 2012 FAOSTAT, http://faostat3.fao.org/home/index.html • Fytogreen: **251** (middle right) • Geography Teachers Association: **61** (bottom)/Source: © Geography Teachers Association of Victoria Inc. Interaction, journal of the GTAV, June 1998. Illustration redrawn by Harry Slaghekke. • Geoscience Australia: **23** (bottom left), **23** (bottom right)/© Commonwealth of Australia Geoscience Australia 2012. This product is released under the Creative Commons Attribution 3.0 Australia Licence. • Getty Images: **211** (bottom)/© egd / Shutterstock.com • Getty Images Australia: **36** (top)/Nick Rains; **52**/Rob Blakers; **56** (top left)/Jon Kopaloff/FilmMagic; **61**/Portland Press Herald; **72** (top right)/ac productions; **72** (middle left)/Cedric Favero; **72** (middle right)/Guy Marks; **73** (middle left)/Nicholas Pitt; **73** (middle right)/Manfred Gottschalk; **84** (top left)/© Mlenny; **87**/Auscape; **87**/George Steinmetz; **88** (top right)/Konrad Wothe / LOOK-foto; **115**/© guenterguni; **158** (top left)/RAUL ARBOLEDA/AFP; **158** (top right)/James P. Blair/National Geographic; **169** (middle)/Anne Clark; **172** (bottom left)/Ben Tweedie/Corbis; **198**/Keren Su; **204** (top b)/PATRICK BAZ / AFP; **204** (top e)/MissHibiscus; **207** (top)/Chris Mellor; **226** (top)/Paul Grand Image; **227** (top)/sot; **252** (middle).d/Steve Debenport; **252** (middle).f/pagadesign; **256**/Kibae Park • iStockphoto: **252** (top).a/Rich Seymour • Judy Mraz: **94** (bottom) • Land Information NZ: **130**/Sourced from Topographic Map 273-09 Egmont. Crown Copyright Reserved. MAPgraphic Pty Ltd, Brisbane • MAPgraphics: **6** (top), **23** (top), **25** (top), **79, 92, 127, 139** (middle), **145** (top), **147, 151** (bottom), **169** (top), **171, 180** (top); **100**/© Geography Teachers Association of Victoria Inc. • Michael Amendolia: **94** (top) • Myriad Editions: **206**/Reproduced with permission from The Atlas of Human Migration by Russell King et al © Myriad Editions | www. myriadeditions.com • NASA: **129**/Image Science and Analysis Laboratory, NASA-Johnson Space Center; **148** (bottom)/ Landsat • NASA Earth Observatory: **200**/Image by Craig Mayhew and Robert Simmon, NASA GSFC; **215, 215**/NASA; **219** (top); **255** • NATSEM: **178** (middle)/AMP-NATSEM Income & Wealth Report Issue 30 "Race against Time" • Newspix: **62** (middle)/Jeff Darmanin; **73** (bottom left)/David Sproule; **145**/Anna Rogers; **244** (top)/Glenn Daniels • Paul F. Downton: **231** (bottom)/Perspective Sketch & Design by Paul F. Downton • Photodisc: **148** (top left), **148** (top right), **160** (bottom) • Public Domain: **70**/Inland Waterway Navigation 2009 Produced by the U.S. Army Corps of Engineers; **175** (top)/United Nations, Department of Economic and Social Affairs, PopulationDivision 2012. World Urbanization Prospects: The 2011 Revision. • RecycleWorks: **66** (top)/Adapted from an image by RecycleWorks www.RecycleWorks.org • Regional Plan Association: **182** (top) • Shutterstock: **2** (top left)/racorn; **2** (top right)/parinyabinsuk; **2** (middle a)/Toa55; **2** (middle b)/l i g h t p o e t; **2** (middle c)/bikeriderlondon; **2** (middle d)/LDprod; **2** (middle e)/SpeedKingz; **2** (middle f)/goodluz; **3** (bottom)/Christian Draghici; **17**/kojihirano; **19, 45**/© Jason Patrick Ross; **19** (top)/Daniel Prudek; **28**/© Sarah Fields Photography; **43**/© tomas del amo; **45**/© Alberto Loyo; **47**/© kukuruxa; **54**/Preto Perola; **56** (top right)/© Peter Clark; **62** (top)/KimPinPhotography; **72** (bottom right)/marzia franceschini; **76**/Sourav and Joyeeta; **78**/Eniko Balogh; **78**/linma; **82**/Roberto Caucino; **83** (top)/© Bill Lawson; **83** (bottom)/© Anderl; **84**/Armin Rose; **84** (top right), **139** (bottom)/AustralianCamera; **88** (top left)/Ralph

Griebenow; **88** (bottom right)/Christopher Halloran; **91** (top right)/N Mrtgh; **99**/steve estvanik; **103**/Dominik Michalek; **107**/© Craig Hanson; **115**/© John A Cameron; **117**/© Francesco Carucci; **117**/© Josemaria Toscano; **121**/NigelSpiers; **127**/ Bjartur Snorrason; **129**/© bartuchna@yahoo.pl; **139** (bottom)/Ashley Whitworth; **139** (bottom)/Paulo Vilela; **149** (bottom)/© Dirk Ercken; **150** (middle left)/Frontpage; **150** (middle right)/© Ammit Jack; **152** (middle right)/YIFANG NIE; **153** (top)/ Dr. Morley Read; **155** (bottom)/Janelle Lugge; **159** (bottom)/Txanbelin; **162** (top)/Dmitry Naumov; **163** (middle)/ Houshmand Rabbani; **164** (top)/Brisbane; **167**/© littlesam; **176** (top)/John-james Gerber; **195** (top)/© Hung Chung Chih; **195** (bottom)/© BartlomiejMagierowski; **200** (bottom)/Ildi Papp; **202**/Tomasz Szymanski; **204** (top a)/Sadik Gulec; **204** (top c)/© Earl D. Walker; **204** (top d)/© Shukaylova Zinaida; **204** (top f)/© forestpath; **204** (top g)/© africa924; **204** (bottom a)/ Daxiao Productions; **204** (bottom b)/© Asianet-Pakistan; **204** (bottom c)/BartlomiejMagierowski; **205** (top d)/© Jane September; **205** (middle left)/Elzbieta Sekowska; **205** (bottom g)/De Visu; **208**/Authentic travel; **215** (bottom)/© fuyu liu; **218** (bottom)/amadeustx; **219** (bottom)/Kzenon; **222** (bottom)/© Andrew Zarivny; **227** (bottom)/© lucarista; **234**/ stocker1970; **236** (bottom)/Johnny Lye; **243**/CTR Photos; **246** (top left)/paintings; **251** (middle left)/Stuart Monk; **252** (top).b/Janelle Lugge; **252** (top).c/THPStock; **252** (middle).e/Blazej Lyjak; **254** (bottom)/tristan tan • Spatial Vision: **6** (bottom), **7** (bottom), **14** (top), **57**, **57**, **72** (middle), **105**, **156** (top up), **168** (bottom), **177** (top), **183**, **187** (top), **201**, **214** (bottom), **240** (bottom); **25** (bottom), **36** (bottom), **40** (bottom), **69**, **90**, **93**, **111**, **114**, **121**, **122**, **123**, **126**, **128**, **173** (top), **186** (bottom), **188** (middle), **206**, **217**, **222** (top), **223** (bottom), **241** (top); **32**/Source: World Map of Carbonate Rock Outcrops v3.0. Map created by Spatial Vision.; **40** (top)/Copyright © 1992 - 2012 UNESCO/World Heritage Centre. All rights reserved. Map created by Spatial Vision.; **174**/Based on UN Map; **212** (top)/Based on data derived from Un-Habitat, SSWm & Spagnoli, Statistics on Poverty, Urbanization and Slums; **218** (top)/Based on UNdata; **235** (bottom)/Copyright Commonwealth of Australia Geoscience Australia 2006. Map drawn by Spatial Vision; **239**/World Bank Data. • Surf Life Saving Australia: **63**/Image supplied by and courtesy of Surf Life Saving Australia • Terry McMeekin: **53** • The World Bank: **72** (top left), **73** (top right)/The World Bank: Average monthly Temperature and Rainfall for Angola Huambo at location -12.43,15.45 from 1990-2012; 72 (bottom left)/The World Bank: Average monthly Temperature and Rainfallfor Botswana at location -19.56,23.27 from 1990-2012http://sdwebx.worldbank.org/climateportal/index.cfm?page=country_historical_ climate&ThisRegion=Africa&ThisCCode=BWA; 73 (bottom right)/The World Bank: Average monthly Temperature and Rainfallfor Australia at location -28.61,137.22 from 1990-2012 • U.S. Geological Survey: **120**; **178** (top)/Landsatlook Viewer • UNEP/GRID-Arendal: 156 (top down)/© Hugo Ahlenius, UNEP/GRID-Arendal http://www.grida.no/graphicslib/ detail/extent-of-deforestation-in-borneo-1950-2005-and-projection-towards-2020_119c • UN-Habitat: **208** (top)/Derived from UN-HABITAT The State of African Cities 2010 • UNSW Science: **73** (top left)/Richard Kingsford UNSW • USGS National Center: **116**, **117**, **120**/US Geological Survey; **126**/US Geological Survey. MAPgraphics Pty Ltd, Brisbane

Text:

• © Australian Curriculum, Assessment and Reporting Authority (**ACARA**) 2010 to present, unless otherwise indicated. This material was downloaded from the Australian Curriculum website (www.australiancurriculum.edu.au) (**Website**) (accessed September 5, 2017) and was not modified. The material is licensed under CC BY 4.0 (https://creativecommons. org/licenses/by/4.0). Version updates are tracked on the 'Curriculum version history' page (www.australiancurriculum.edu. au/Home/CurriculumHistory) of the Australian Curriculum website. • Australian Bureau of Statistic: **187** (bottom)/Based on Australian Bureau of Statistics data Census of Population and Housing 2011 © Commonwealth of Australia , licensed under a Creative Commons Attribution 2.5 Australia licence; **187** (middle)/Source: Australian Bureau of Statistics, licensed under a Creative Commons Attribution 2.5 Australia licence • Creative Commons: 177 (bottom)/© Infrastructure Australia 2015; **188** (top)/© Commonwealth of Australia • FSC International Center GmbH: **160** (middle) • International Association of Antarctica Tour Operators: **101** • John Wiley & Sons Australia, Ltd: **138** (bottom) • U.S. Census Bureau: **223** (middle) • United Nations: **170** (top)/Data obtained from United Nations, Department of Economic and Social Affairs, Population Division 2015. World Population Prospects: The 2015 Revision, p 1 • United Nations Population Div.: **238**

Every effort has been made to trace the ownership of copyright material. Information that will enable the publisher to rectify any error or omission in subsequent reprints will be welcome. In such cases, please contact the Permissions Section of John Wiley & Sons Australia, Ltd.

TOPIC 1
The world of Geography

1.1 Overview

Numerous **videos** and **interactivities** are embedded just where you need them, at the point of learning, in your learnON title at www.jacplus.com.au. They will help you to learn the content and concepts covered in this topic.

1.1.1 Work and careers in Geography

As a student of Geography, you are starting to build the knowledge and skills that will be needed by you and your community now and into the future. The concepts and skills that you will use will not only help you in Geography but they can also be applied to everyday situations, such as finding your way from one place to another. Studying Geography may even help you in a future career here in Australia or somewhere overseas.

Throughout the year you will be studying topics that will give you a better understanding of the world around you — both the local and global environment. You will be investigating issues that need to be addressed now and also options for the future.

Explore: skills you need for a job

Many questions come up during a typical Geography class, such as the ones below in table 1. These questions need to be answered in the real world by people in a wide variety of occupations. They all have links with Geography.

TABLE 1 Examples of occupations that use Geography

Question	Occupations/organisations that try to answer these questions
• How high *is* Mount Everest? How do we know?	Surveyor, Cartographer
• How can we protect our parks and wildlife?	Park ranger, Planner, Environmental manager
• Where should we establish a new suburb for our future population?	Urban planner, Demographer
• How can we prepare for future droughts and floods?	Civil engineer
• Does our town really have enough water? • Where should we build a new dam? Should we build a new dam?	Coastal engineer, Hydrologist, Cartographer
• Should a boat marina be built at location X or at location Y?	Oceanographer
• Do we have good quality drinking water?	Chemist, Hydrologist
• How do countries such as India and China deal with their air pollution problems?	Environmental scientist/Manager
• How do we provide aid to other countries?	Air Force, Navy, Army Officer. Red Cross, World Vision and other aid agencies.
• How do we build sustainable housing?	Architect, Landscape architect, Civil engineer/ Construction manager, Town planner, Real estate salesperson

Think: who are you and what is your position in the world?

Do you know much about the occupations mentioned in table 1? Are any of interest to you?

The first step in thinking about your future is to consider questions such as:

- Who am I?
- What are my interests?
- What do I enjoy doing?
- What am I good at?
- When I leave school I would like to …

1.1.2 Geography careers on the move

A great part of studying Geography is being able to explore the many occupations and areas that it opens up. In table 2 are some occupations that you may not have thought studying Geography could lead you into.

TABLE 2 Would I enjoy …

… working indoors?	… working outdoors?	… helping people?

• Land economist • Landscape designer • Real estate salesperson • Geoscience technician • Travel consultant	• Surveyor • Mining engineer • Geologist • Landscape architect • Cartographer	• Park ranger • Paramedic • Navy officer • Fireman • Tour guide
… designing new places to live?	… improving people's wellbeing?	… doing research?

• Urban planner • Architect • Landscape architect • Horticulturist	• Natural resource manager • Demographer	• Meteorologist • Anthropologist • Geophysicist • Hydrographer • Environmental scientist

1.1.3 Finding my way as a local and global citizen

A wide range of exciting new jobs are developing in the spatial sciences which use geographical tools such as GPS, GIS, satellite imaging and surveying. These tools help people make important decisions about managing and planning places and resources. Whether it be how to manage water somewhere in the Middle East or how best to design a new housing estate here in Australia, these skills and occupations will be an important part of working as a global citizen.

1.1 Activities

To answer questions online and to receive **immediate feedback** and **sample responses** for every question, go to your learnON title at www.jacplus.com.au. *Note:* Question numbers may vary slightly.

Predict

1. Consider spatial technologies and work and enterprise careers of the future. Which geographical tools do you predict will be used by:
   - a meteorologist
   - a naval officer
   - an airline pilot
   - a farmer?

Discover

2. We can develop a better understanding of work and enterprise by exploring what others have to say about their careers. Use the **Geocareers** weblink in the Resources tab to help you locate one male and one female geographer working as local or global citizens.
   - What career path did they follow?
   - How did they include their passion for geography into their career journey?
   - What advice did they share about their career journey?

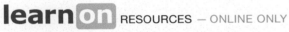

RESOURCES — ONLINE ONLY

Explore more with this weblink: Geocareers

1.2 Geographical concepts

1.2.1 Overview

Geographical concepts help you to make sense of your world. By using these concepts you can both investigate and understand the world you live in, and you can use them to try to imagine a different world. The concepts help you to think geographically. There are seven major concepts: *place*, *space*, *environment*, *interconnection*, *sustainability*, *scale* and *change*.

In this book, you will use the seven concepts to investigate two units: *Landforms and landscapes* and *Changing nations*.

FIGURE 1 A way to remember these seven concepts is to think of the term SPICESS.

Space
Place
Interconnection
Change
Environment
Sustainability
Scale

1.2.2 What is space?

Everything has a location on the space that is the surface of the Earth, and studying the effects of location, the distribution of things across this space, and how the space is organised and managed by people, helps us to understand why the world is like it is.

A place can be described by its absolute location (latitude and longitude) or its relative location (in what direction and how far it is from another place).

FIGURE 2 The distribution of the world's deserts

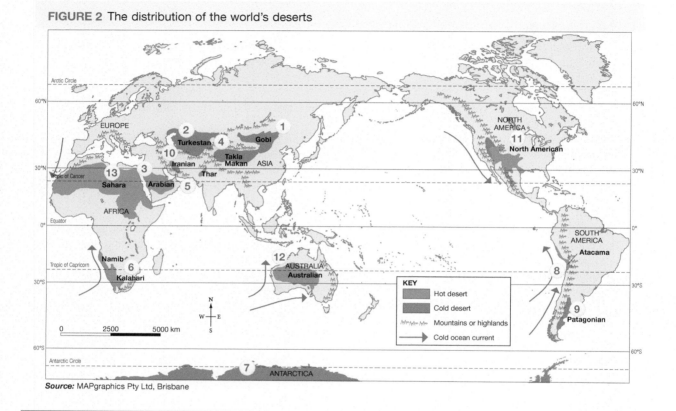

Source: MAPgraphics Pty Ltd, Brisbane

1.2.2 Activities

To answer questions online and to receive **immediate feedback** and **sample responses** for every question, go to your learnON title at www.jacplus.com.au. *Note:* Question numbers may vary slightly.

Refer to figure 2.

1. Use an atlas to give the absolute location (latitude and longitude) of Mecca, in the Arabian Desert. What is the relative location of Mecca from Australia?
2. Describe the *spatial* distribution of the world's deserts in relation to the tropics.
3. In what direction and approximately how far is the Thar Desert from the Arabian Desert; the Atacama Desert and the Namib Desert?
4. How is the location of the Namib Desert influenced by cold ocean currents?
5. Look at a population map of Australia (see '1.2.4 What is Interconnection?', figure 5). What is the relationship over *space* between the location of Australia's deserts and the location of Australia's population?

  Deepen your understanding of this topic with related case studies and questions.
◗ **Space**

1.2.3 What is place?

The world is made up of places, so to understand our world we need to understand its places by studying their variety, how they influence our lives and how we create and change them.

You often have mental images and perceptions of places — rich and poor cities, suburbs, towns or neighbourhoods — and these may be very different from someone else's perceptions of the same places.

FIGURE 3 The Paraisópolis favela (slum), home to 60 000 people, is situated next to the gated complexes of the wealthy Morumbi district of São Paulo.

1.2.3 Activities

To answer questions online and to receive **immediate feedback** and **sample responses** for every question, go to your learnON title at www.jacplus.com.au. *Note:* Question numbers may vary slightly.

Refer to figure 3.
1. Where in the world is this *place* located?
2. What effects have people had on this *place*?
3. List the differences you observe in the way people live on each side of this settlement.
4. How is this *place* similar to or different from the *place* where you live?
5. What decisions could be made to improve or *change* this *place*?
6. How might the *environment* of this *place* affect the people who live there?
7. How does the *place* where these people live affect their lives?

 Deepen your understanding of this topic with related case studies and questions.
❯ **Place**

1.2.4 What is interconnection?

People and things are connected to other people and things in their own and other places, and understanding these connections helps us to understand how and why places are changing.

An event in one location can lead to change in a place some distance away.

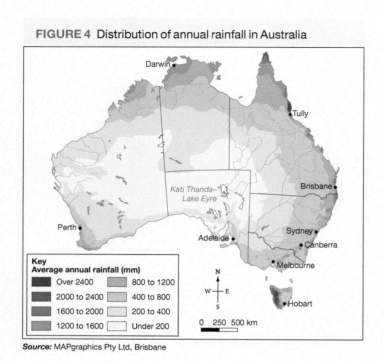

FIGURE 4 Distribution of annual rainfall in Australia

Source: MAPgraphics Pty Ltd, Brisbane

FIGURE 5 Australia's population distribution

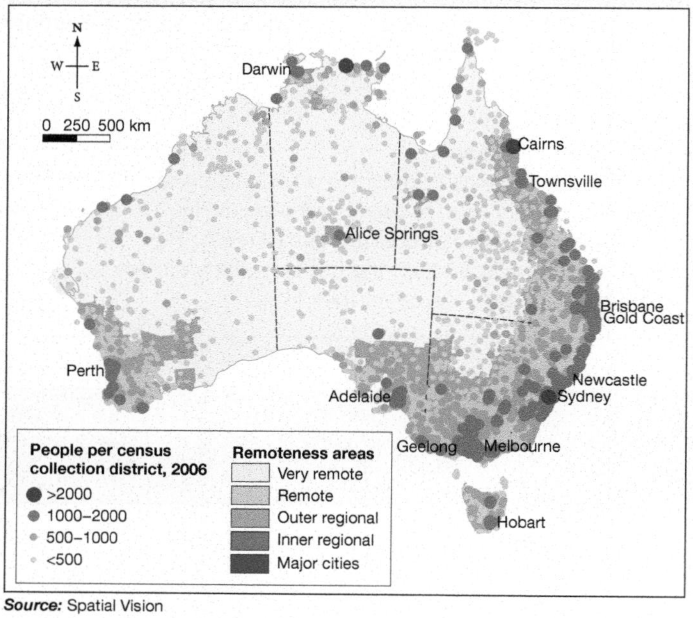

Source: Spatial Vision

1.2.4 Activities

To answer questions online and to receive **immediate feedback** and **sample responses** for every question, go to your learnON title at www.jacplus.com.au. *Note:* Question numbers may vary slightly.

Refer to figures 4 and 5.

1. Look at the rainfall map. What is the relationship, or *interconnection*, between the distribution of rivers and the distribution of rainfall?
2. In small groups, think of ways in which the use of water upstream in these rivers could affect *environments* and people downstream.
3. Write a summary statement that describes the *interconnection* between rainfall in Australia and where people live.
4. Look at an atlas map showing the distribution of vegetation in Australia. How are rainfall and vegetation *interconnected*?
5. Look at an atlas map. What is the *interconnection* between rainfall distribution and Australia's population density (number of people per census district)?
6. How might these *interconnections* affect how people live?

 my**World**Atlas   Deepen your understanding of this topic with related case studies and questions.
⊙ Interconnection

1.2.5 What is change?

The concept of change is about using time to better understand a place, an environment, a spatial pattern or a geographical problem.

Some changes can be fast and easily observed, but others are very slow. Cities, for example, can expand outwards over a number of years. Similarly, landforms generally change very slowly, as with the formation of mountains. But some landscape change can be very fast, as is the case with landslides, volcanic eruptions and deforestation.

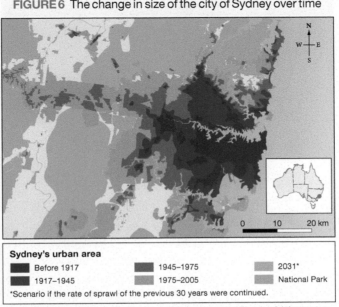

FIGURE 6 The change in size of the city of Sydney over time

Sydney's urban area

- Before 1917
- 1917–1945
- 1945–1975
- 1975–2005
- 2031*
- National Park

*Scenario if the rate of sprawl of the previous 30 years were continued.

Source: Provided by Metropolitan Strategy, NSW Department of Planning & Infrastructure.

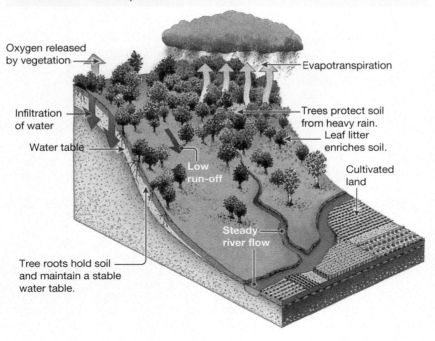

FIGURE 7a Landscape before deforestation

Oxygen released by vegetation

Evapotranspiration

Infiltration of water

Trees protect soil from heavy rain.

Leaf litter enriches soil.

Water table

Low run-off

Cultivated land

Steady river flow

Tree roots hold soil and maintain a stable water table.

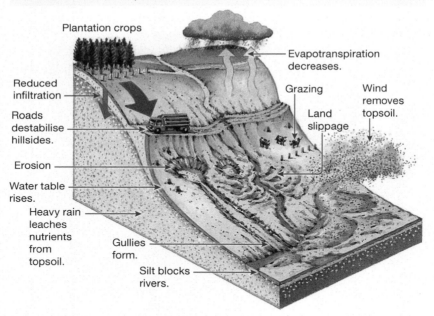

FIGURE 7b Landscape after deforestation

Plantation crops

Evapotranspiration decreases.

Reduced infiltration

Grazing

Wind removes topsoil.

Roads destabilise hillsides.

Land slippage

Erosion

Water table rises.

Heavy rain leaches nutrients from topsoil.

Gullies form.

Silt blocks rivers.

1.2.5 Activities

To answer questions online and to receive **immediate feedback** and **sample responses** for every question, go to your learnON title at www.jacplus.com.au. *Note:* Question numbers may vary slightly.

Refer to figure 6.

1. How has Sydney *changed* over time? How long has it taken for the city to spread to the furthest areas shown on the map?

2. What main natural feature attracted the earliest settlement?

3. What impact do you think this growth has had on the natural *environment*?
4. What technological *changes* in transport have allowed Sydney to spread and grow over time?
Refer to figures 7a and 7b.
5. List the *changes* that would have caused the slippage to occur.
6. What *interconnections* are there between:
   (a) vegetation cover and soil stability
   (b) vegetation cover and infiltration
   (c) high run-off and erosion?
7. List all the effects of the landslide on people and the *environment*.
8. Write a summary statement about the pace of *change* and the impact on people and the *environment* in these two examples.

myWorldAtlas   Deepen your understanding of this topic with related case studies and questions.
   ❯ Change

1.2.6 What is environment?

People live in and depend on the environment, so it has an important influence on our lives.

The environment, defined as the physical and biological world around us, supports and enriches human and other life by providing raw materials and food, absorbing and recycling wastes, and being a source of enjoyment and inspiration to people.

FIGURE 8 Uranium mining in Colorado, United States. Many deserts contain valuable mineral deposits.

1.2.6 Activities

To answer questions online and to receive **immediate feedback** and **sample responses** for every question, go to your learnON title at www.jacplus.com.au. *Note:* Question numbers may vary slightly.

Refer to figure 8.

1. Describe the *environment* in which this mine is located. What evidence is there that this mine is located close to a river?
2. Is this a natural or a human *environment*? Justify your choice.
3. What impact might this mine have on the surrounding *environment*?
4. What resource or raw material is mined here?
5. How might the local and national economies benefit from what is happening in this *environment*?
6. How has technology helped *change* this *environment*?
7. In your opinion, what are the positive and negative aspects of the way this *environment* has been used?

1.2.7 What is sustainability?

Sustainability is about maintaining the capacity of the environment to support our lives and those of other living creatures.

Sustainability is about the interconnection between the human and natural world and who gets which resources and where, in relation to conservation of these resources and prevention of environmental damage.

FIGURE 9 The Vatican is the world's smallest independent state and is hoping to become the first solar-powered nation in the world. It plans to create Europe's largest solar power plant, which will provide enough energy to power all of the state's 40 000 households. The roof of the Paul VI Hall is now covered in photoelectric cells.

1.2.7 Activities

To answer questions online and to receive **immediate feedback** and **sample responses** for every question, go to your learnON title at www.jacplus.com.au. *Note:* Question numbers may vary slightly.

Refer to figure 9.

1. What resource is being harnessed by the solar panels? Brainstorm a list of all the ways in which this energy can be used.
2. How is using solar panels an example of *sustainable* energy use? The use of what other resource will be reduced by using these solar panels?
3. Using the caption, outline the *sustainability* interests and values that the Vatican City is following by becoming a solar city.
4. What are some examples of other types of *sustainable* energy? Make a list of these with a partner.

myWorldAtlas   Deepen your understanding of this topic with related case studies and questions.
❯ Environment
❯ Sustainability

1.2.8 What is scale?

When we examine geographical questions at different spatial levels we are using the concept of scale to find more complete answers.

Scale can be from personal and local to regional, national or global. Looking at things at a range of scales allows a deeper understanding of geographical issues.

FIGURE 10 Perth, Western Australia. Building sustainable communities means we have to work at various scales.

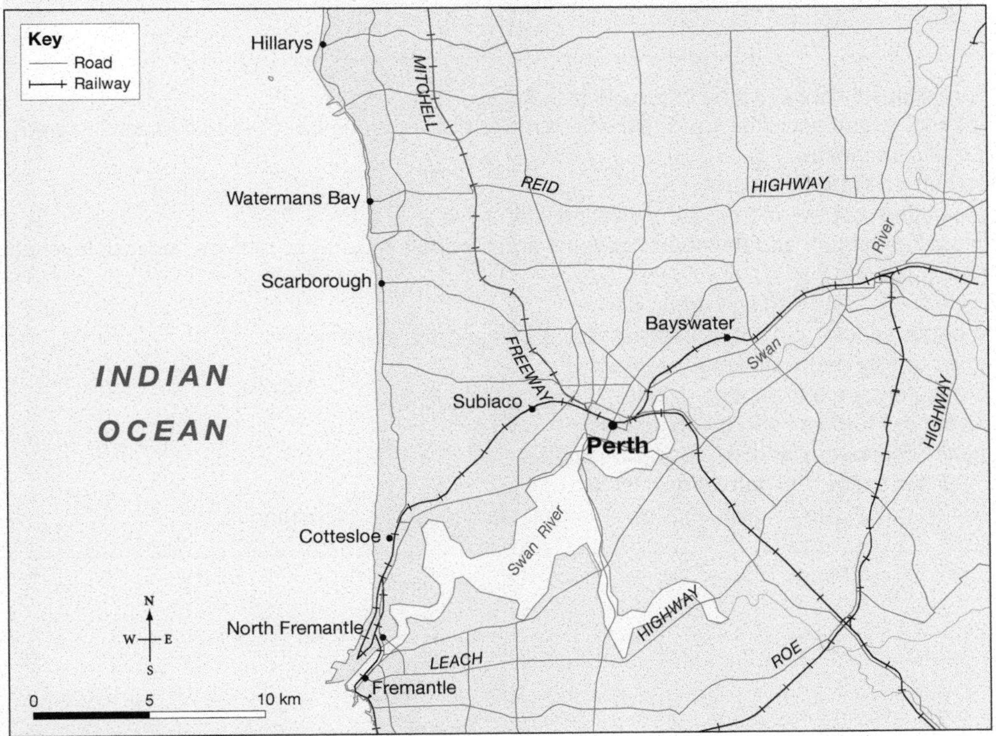

Key
— Road
⊢—+ Railway

Hillarys
MITCHELL
REID
HIGHWAY
River
Watermans Bay
Scarborough
Bayswater
Swan
INDIAN
FREEWAY
Subiaco
OCEAN
Perth
HIGHWAY
Swan River
Cottesloe
HIGHWAY
N
W   E
S
North Fremantle
ROE
LEACH
0   5   10 km
Fremantle

Source: © Commonwealth of Australia Geoscience Australia 2006.

FIGURE 11a An urban organic community in Perth

FIGURE 11b An aerial view of the Swan River and the city of Perth

Ways to improve sustainability at the local scale include:
- reducing the ecological footprint
- protecting the natural environment
- increasing community wellbeing and pride in the local area
- changing behaviour patterns by providing better local options
- encouraging compact or dense living
- providing easy access to work, play and schools.

Ways to improve sustainability at the city scale include:
- building strong central activities areas (either one major hub, or a number of specified activity areas)
- reducing traffic congestion
- protecting natural systems
- avoiding suburban sprawl and reducing inefficient land use
- distributing infrastructure and transport networks equally and efficiently to provide accessible, cheap transportation options
- promoting inclusive planning and urban design
- providing better access to healthy lifestyles (e.g. cycle and walking paths)
- improving air quality and waste management
- using stormwater more efficiently
- increasing access to parks and green spaces
- reducing car dependency and increasing walkability
- promoting green space and recreational areas
- demonstrating a high mix of uses (e.g. commercial, residential and recreational).

1.2.8 Activities

To answer questions online and to receive **immediate feedback** and **sample responses** for every question, go to your learnON title at www.jacplus.com.au. *Note:* Question numbers may vary slightly.

Refer to figure 10.
1. What main information is this map trying to show?

Refer to figures 11a and 11b.
2. The photo in figure 11a shows a zoomed-in view of the Perth city area. Do you see more or less detail at the zoomed-in *scale*?
3. Identify two city aims listed above that could not be implemented at the local *scale*.
4. Identify one local or neighbourhood aim listed above that could not be implemented at the city *scale*.
5. Describe any local action where you live that tries to improve *sustainability*. You could talk to your parents about this, or contact your local council to see what they are trying to do about the issue.

my**World**Atlas Deepen your understanding of this topic with related case studies and questions.
> Scale

1.3 Review

1.3.1 Applying the concepts

For the past 7000 years, the sandy beaches close to Adelaide in South Australia have been eroding and the sand has been moved northwards by the prevailing winds from the south-west. This has resulted in the growth of a peninsula and large dune system at North Haven.

As many people have now settled along this coastline, the sand movement is being managed to stop erosion by trapping the sand or trucking and piping it back to the south. The cost of this action can be very high.

FIGURE 1 Piping sand from north to south along Adelaide's beaches

FIGURE 2 The movement of sand northwards along the Adelaide Metropolitan coastline

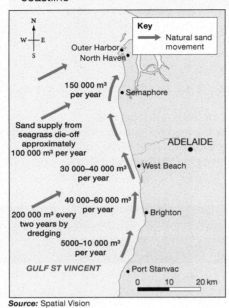

Source: Spatial Vision

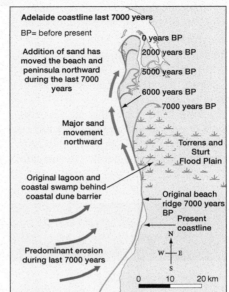

Source: Spatial Vision

1.3.1 Activities

To answer questions online and to receive **immediate feedback** and **sample responses** for every question, go to your learnON title at www.jacplus.com.au. *Note:* Question numbers may vary slightly.

1. Use an atlas to describe the location of Adelaide relative to where you live. What is its absolute location? (*space*)
2. Is the Adelaide coast a *place* that you would like to live? Why or why not?
3. How would the perception of coastal erosion differ between people who live in coastal settlements and those living five or more kilometres inland? (*place*)
4. What are the human natural characteristics of this *place*? Do you think this coastal *environment* is natural or human? Justify your answer.
5. What is the length of the coastline shown on the maps in figure 2? (*scale*)
6. Describe how this *environment* may have looked 7000 years ago. How has it *changed* over time?
7. How have people *changed* this *environment*?
8. What is the relationship (*interconnection*) between:
   (a) prevailing wind and sand movement (the prevailing wind along this coastline comes from the south-west)
   (b) sand movement and formation of coastal features?
9. Which local management *scale* is shown in figure 1?
10. How has technology helped manage this coastline? (*change*)
11. What *changes* might occur if the sand movement was not managed?
12. Is the movement of sand to stop beach erosion a *sustainable* activity — that is, is it worth stopping the movement of sand? Explain your answer.

UNIT 1
LANDFORMS AND LANDSCAPES

Have you ever stood on a hill, or high ground, and looked at the scenery and landscape in front of you? From a height you can see a variety of different landforms such as mountains, valleys and plains. So, how are different landforms actually created? And what causes the hazards we need to deal with?

View across the Dolomite mountain range, Italy

TOPIC 2
Introducing landforms and landscapes

2.1 Overview

Numerous **videos** and **interactivities** are embedded just where you need them, at the point of learning, in your learnON title at www.jacplus.com.au. They will help you to learn the content and concepts covered in this topic.

2.1.1 Introduction

Landscapes are the visible features of the land, ranging from the icy landscapes of polar regions and lofty mountain ranges, through to forests, deserts and coastal plains. Shaped by physical processes over millions of years, they have been overlaid by the presence of humans.

Landforms in a desert landscape: the Colorado River meanders through Horseshoe Bend in Arizona, United States.

Starter questions

1. Describe key features of the landscape shown in the image of Horseshoe Bend on the previous page.
2. What do you think might be the difference between a natural and a human *environment*?
3. Brainstorm as a class different landscapes around the world. Describe your favourite landscape and explain why it appeals to you.

2.2 How are landscapes different?

2.2.1 Types of landscapes

There are many different landscapes across the Earth, and similarities can be observed within regions. Variations in landscapes are influenced by factors such as climate; geographical features, including mountains and rivers; latitude; the impact of humans; and where the landscapes are located.

FIGURE 1 Selected world landscapes

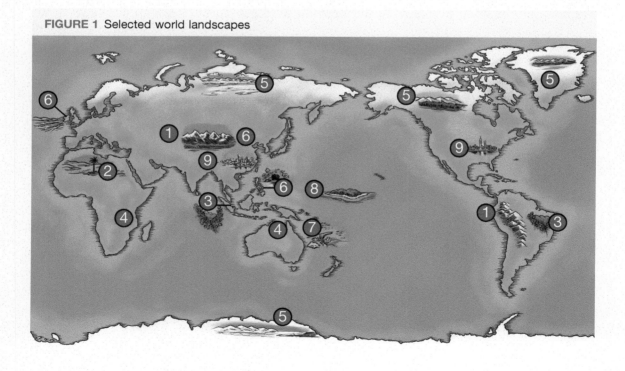

1 Mountains

Mountains rise above the surrounding landscape. They often have steep sides and high peaks and are the result of processes operating deep inside the Earth. Some reach high into the atmosphere where it is so cold that snow is found on their peaks.

2 Deserts

Deserts are areas of low rainfall; they are an arid or dry environment. They can experience temperature extremes: hot by day and freezing at night. However, not all deserts are hot. Antarctica is the world's largest desert, and the Gobi Desert, located on a high **plateau** in Asia, is also a cold desert.

3 Rainforests

Rainforests are the most diverse landscapes on Earth. They are found in a variety of climates, ranging from the hot wet tropics to the cooler temperate areas. The lush vegetation found in these regions depends on a high level of rainfall. Over 50 per cent of all known plant and animal species are found within them. In addition, many of our foods and medicines come from rainforests.

4 Grasslands

Grasslands, or savanna, are sometimes seen as a transitional landscape found between forests and deserts. They contain grasses of varying heights and coarseness, and small or widely spaced trees. They are often inhabited by grazing animals.

5 Polar regions

Polar regions and tundra can be found in polar and alpine regions. Characterised by **permafrost**, they are too cold for trees to grow. Vegetation such as dwarf shrubs, grasses and lichens have adapted to the extreme cold and short growing season. **Glaciers** often carve spectacular landscape features.

6 Karst landscapes

Karst landscapes form when mildly acidic water flows over soluble rock such as limestone. Small fractures form, which increase in size over time and lead to underground drainage systems developing. Common landforms include limestone pavements, disappearing rivers, reappearing springs, sinkholes, caves and karst mountains. Around 25 per cent of the world's population obtains water from karst **aquifers**.

7 Aquatic landscapes

Aquatic landscapes cover around three-quarters of the Earth and can be classified as freshwater or marine. Marine landscapes are the saltwater regions of the world, and include oceans and coral reefs. Freshwater landscapes are found on land, and include lakes, rivers and wetlands.

8 Islands

Islands are areas of land that are completely surrounded by water. They can be continental or oceanic. Continental islands lie on a continental shelf — an extension of a continent that is submerged beneath the sea. Oceanic islands rise from the ocean floor and are generally volcanic in origin. A group or chain of islands is known as an archipelago.

FIGURE 2 At 8848 metres, Mount Everest in the Himalayas is the highest mountain on Earth.

FIGURE 3 These karst caves in southern China are protected.

⑨ Built landscapes

Human or built landscapes are those that have been altered or created by humans.

2.2 Activities

To answer questions online and to receive **immediate feedback** and **sample responses** for every question, go to your learnON title at www.jacplus.com.au. *Note:* Question numbers may vary slightly.

Remember

1. Explain the difference between the terms *natural* and *built environment*.
2. What factors make landscapes different?
3. List as many different human or built *environments* as you can think of.

Explain

4. Why do you think people *change* landscapes?

Discover

5. The map in figure 1 shows the wide variety of landscapes found on the surface of the Earth. However, it does not show all locations for each landscape type. Investigate one of the featured landscapes and find out other *places* in which it is found. Show this information on a map. Annotate your map with information from this subtopic and characteristics of your landscape.

Think

6. Copy the following table into your workbook.

Characteristics	How people use it	Positive impacts	Negative impacts

   (a) Select one of the landscape types described in this subtopic and complete the table.
   (b) Which list is larger — the positive impacts or negative impacts?
   (c) Review the column of negative impacts. Select three of these impacts and suggest a way in which the *environment* could be used in a more *sustainable* way.
7. Describe how the *scale* of the following landscapes might differ around the world: deserts, polar regions, aquatic landscapes and islands.

 RESOURCES — ONLINE ONLY

 Try out this interactivity: Landscapes galore (int-3102)

 Deepen your understanding of this topic with related case studies and questions.
❯ **Grasslands**

2.3 What landforms make up Australia?

2.3.1 What processes have shaped Australasia?

The tectonic forces of folding, faulting and volcanic activity have created many of Australia's major landforms. Other forces that work on the surface of Australia, and give our landforms their present appearance, are **weathering**, mass movement, **erosion** and **deposition**.

Australia is an ancient landmass. The Earth is about 4600 million years old, and parts of the Australian continent are about 4300 million years old.

Over millions of years, Australia has undergone many changes. Mountain ranges and seas have come and gone. As mountain ranges eroded, sediments many kilometres thick were laid down over vast areas. These sedimentary rocks were then subjected to folding, faulting and uplifting. This means that the rocks that make up the Earth's crust have buckled and folded along areas of weakness, known as faults. Sometimes, fractures or breaks occur, and forces deep within the Earth cause sections to be raised, or uplifted. Over time the forces of weathering and erosion have worn these down again. Erosion acts more quickly on softer rocks, forming valleys and bays. Harder rocks remain as mountains, hills and coastal headlands.

Because it is located in the centre of a **tectonic plate**, rather than at the edge of one, Australia currently has no active volcanoes on its mainland, and has very little tectonic lift from below. This means its raised landforms such as mountains have been exposed to weathering forces for longer than mountains on other continents and are therefore more worn down.

About 33 million years ago, when Australia was drifting northwards after splitting from Antarctica, the continent passed over a large **hotspot**. Over the next 27 million years, about 30 volcanoes erupted while they were over the hotspot. The oldest eruption was 35 million years ago at Cape Hillsborough, in Queensland, and the most recent was at Macedon in Victoria around six million years ago. Over millions of years, these eruptions formed a chain of volcanoes in eastern and south-eastern Australia, that are known today as the Great

FIGURE 1 Many of Queensland's mountain peaks were formed by volcanic activity around 20 million years ago. The Glasshouse Mountains, north of Brisbane, are volcanic plugs. They are composed of volcanic rock that hardened in the vent of a volcano. Over millions of years, weathering and erosion have worn away the softer rock that surrounded the vent, leaving only the plugs.

Dividing Range (see figure 2). At present, the hotspot that caused the earlier eruptions is probably beneath Bass Strait.

The present topography of much of Australia results from erosion caused by ice. For example, about 290 million years ago a huge icecap covered parts of Australia. After the ice melted, parts of the continent subsided and were covered by **sediment**, forming sedimentary basins (a low area where sediments accumulate) such as the Great Artesian Basin. On a smaller scale, parts of the Australian Alps and Tasmania have also been eroded by glaciers during the last ice age.

Rivers and streams are another cause of erosion, having carved many of the valleys in Australia's higher regions.

When streams, glaciers and winds slow down, they deposit or drop the material they have been carrying. This is called deposition. Many broad coastal and low-lying inland valleys have been created by stream deposition. These areas are called floodplains.

2.3.2 What are Australia's landform regions?

The topography of Australia can be divided into four major regions (see figure 3).

- The coastal lowlands around Australia's edge are narrow and fragmented. The plains often take the form of river valleys, such as the Hunter River Valley near Newcastle.
- The eastern highlands region (which includes the Great Dividing Range) is mainly a series of tablelands and plateaus. Most of the area is very rugged, because rivers have cut deep valleys. It is the source of most of Australia's largest rivers, including the Fitzroy, Darling and Murray. The highest part is in the south-east, where a small alpine area is snow-covered for more than half the year.
- The central lowlands are a vast area of very flat, low-lying land that contains three large **drainage basins**: the Carpentaria Lowlands in the north, the Lake Eyre Basin in the centre (see figure 4) and the Murray–Darling Basin in the south.
- The Great Western Plateau is a huge area of tablelands, most of which are about 500 metres above sea level. It includes areas of gibber (or stony) desert and sandy desert. There are several rugged upland areas, including the Kimberley and the McDonnell Ranges.

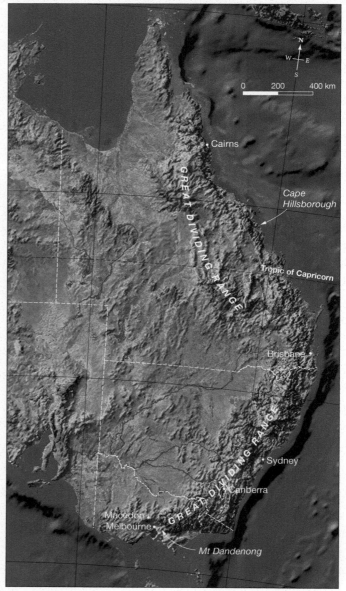

FIGURE 2 Relief map of Australia's east coast. The Great Dividing Range stretches from north of Cairns in Queensland to Mount Dandenong near Melbourne in the south.

Source: Spatial Vision

FIGURE 3 Australia's four major landform regions

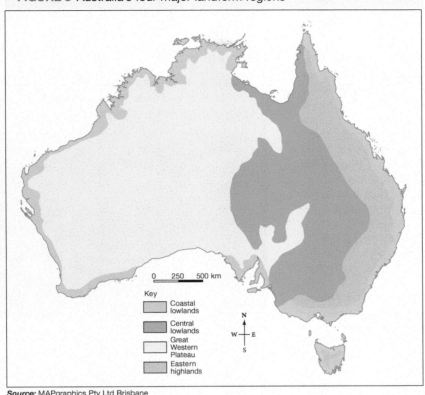

Key
Coastal lowlands
Central lowlands
Great Western Plateau
Eastern highlands

0   250   500 km

N
W—E
S

Source: MAPgraphics Pty Ltd Brisbane

FIGURE 4 Kati Thanda–Lake Eyre, the lowest point on the Australian mainland, is part of the Great Artesian Basin. It is 15 metres below sea level. Once a freshwater lake, the region is now the world's largest salt pan. The evaporated salt crust shows white in the satellite image (a) left. The lake fills with water only three or four times each century, transforming it into a haven for wildlife. Deep water is shown as black in image (b) right.

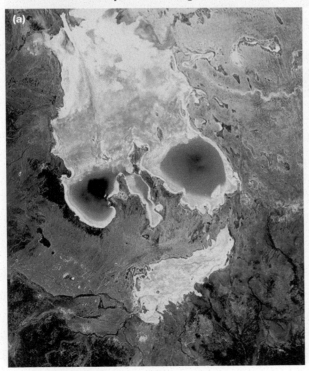

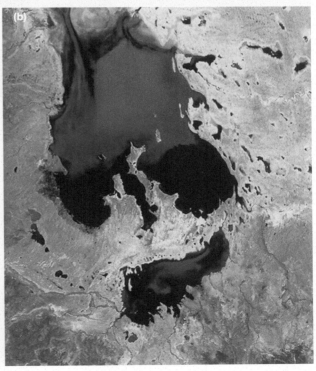

2.3.3 The Murray–Darling Basin

CASE STUDY

Water issues in the Murray–Darling Basin region

The Murray–Darling Basin covers about one million square kilometres, and more than 20 major rivers flow into it. It has a wide variety of landscapes, ranging from alpine areas in the south-east to plains in the west. The basin produces 43 per cent of Australia's food and over 40 per cent of Australia's total agricultural income.

The Murray–Darling Basin is the largest and most important drainage basin in Australia, covering one-seventh of the continent. However, the amount of water flowing through it in one year is about the same as the *daily* flow of the Amazon River.

The basin is facing severe problems.

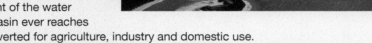

FIGURE 5 Aerial view of the Murray River, where it enters the Coorong and Lake Alexandrina in South Australia

- Only about 20 per cent of the water flowing through the basin ever reaches the sea. The rest is diverted for agriculture, industry and domestic use.
- The Murray supplies about 40 per cent of Adelaide's drinking water. The quality of the water continues to decline, mainly because of salinity levels.
- Approximately 50 to 80 per cent of the wetlands in the basin have been severely damaged or destroyed, and more than a third of the native fish species are threatened with extinction.
- In 2008, inflows into the river system were at their lowest levels since records began 117 years earlier.
- An estimate of weather trends shows that the flow to the Murray River mouth may be reduced by a further 25 per cent by 2030. However, with the added problem of climate change, it is predicted that precipitation in the Murray–Darling catchment will decrease, so that the reduction in flow to the mouth could be as high as 70 per cent.

Deepen your understanding of this topic with related case studies and questions.
❯ **Murray–Darling Basin**

2.3.4 How does water flow across the land?

Permanent rivers and streams flow in only a small proportion of the Australian continent. Australia is in fact the driest of all the world's inhabited continents. It has:

- the least amount of run-off
- the lowest percentage of rainfall as run-off
- the least amount of water in rivers
- the smallest area of permanent wetlands
- the most variable rainfall and stream flow.

Australia has many lakes, but they hold little water compared with those found on other continents. The largest lakes are Kati Thanda–Lake Eyre (see figures 6 and 7) and Lake Torrens in South Australia. During the dry seasons, these become beds of salt and mud. Yet an inland sea did once exist in this area. It covered about 100 000 square kilometres around present-day Kati Thanda–Lake Eyre and Lake Frome. South Australia is Australia's driest state, and has very few permanent rivers and streams.

FIGURE 6 Australia's drainage basins

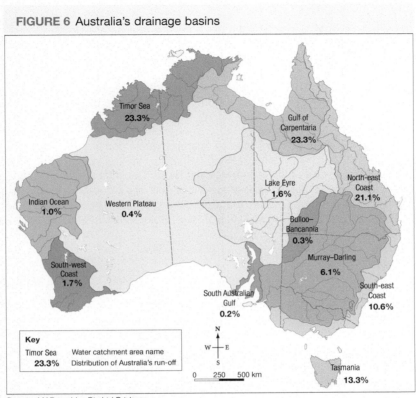

Source: MAPgraphics Pty Ltd Brisbane

FIGURE 7 Kati Thanda–Lake Eyre and surrounding drainage systems

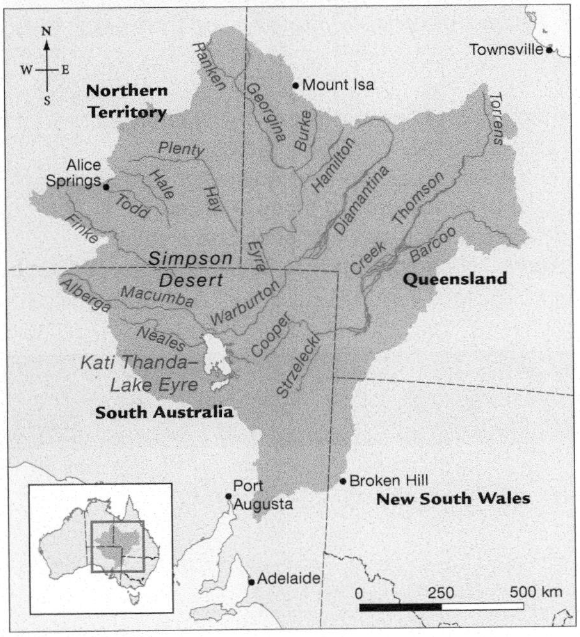

Source: Spatial Vision

2.3 Activities

To answer questions online and to receive **immediate feedback** and **sample responses** for every question, go to your learnON title at www.jacplus.com.au. *Note:* Question numbers may vary slightly.

Remember

1. In your own words, explain what is meant by the terms *folding, faulting* and *uplift*.
2. Describe some of the physical *changes* Australia's landmass has undergone.

Explain

3. Describe the major characteristics of Australia's four main landform regions.
4. Explain why Australia is so low in altitude and flat compared with other continents.

Discover

5. Use your atlas to list the highest mountains in each Australian state and territory. Describe the location of each.
6. Use Google Earth to view any part of the Murray–Darling Basin. Describe the landscape that you see.
7. Divide your class into four groups. Assign each group one of Australia's landform regions to investigate. Collectively compile a list of landforms that are found in each region. Then have each member of the group investigate a different landform and prepare a series of PowerPoint slides that show the following:
   (a) the landform
   (b) where it is located
   (c) how it was formed
   (d) whether people might want to visit this landform, including the reasons why it is or is not a popular landform.
   Put the individual presentations together for viewing by the rest of the class.

Predict

8. Use your atlas to find the Cape Hillsborough and Macedon volcanoes, or refer to figure 2.
   (a) Calculate the distance between them.
   (b) Use the information in this subtopic to work out the rate at which the Australian landmass is moving.
   (c) How far has Australia moved over the Bass Strait hotspot? Now calculate where under Bass Strait this hotspot might now lie.
   (d) Use the information in this subtopic to explain why this hotspot has *changed* its location over time.
9. It is said that the amount of water that flows down the Amazon River in a day is more than flows down the Murray in a year.
   (a) What does that tell you about how dry Australia's climate is?
   (b) How might this affect the way the *environment* around the Murray River is affected?

Think

10. Australia is an ancient landmass and has undergone many *changes* over millions of years. In groups, brainstorm and compile lists under the following headings.
    • Physical *changes* that have taken place on the Australian landmass
    • Tectonic processes that have contributed to these *changes*
    • *Changes* caused by processes such as weathering and erosion
    Within your group, write a series of paragraphs that explain the *interconnection* between these factors.

Deepen your understanding of this topic with related case studies and questions.
➤ **Uluru**

2.4 SkillBuilder: Recognising land features

WHAT ARE LAND FEATURES?

Land features are landforms with distinct shapes, such as hills, valleys and mountains. You can recognise these as you look around your natural environment. On topographic maps you can recognise land features from the patterns formed by the contour lines.

Go online to access:

- a clear step-by-step explanation to help you master the skill
- a model of what you are aiming for
- a checklist of key aspects of the skill
- a series of questions to help you apply the skill and to check your understanding.

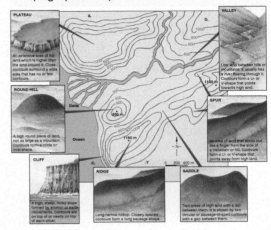

FIGURE 1 Landforms matched to a topographic map

 RESOURCES — ONLINE ONLY

Watch this eLesson: Recognising land features (eles-1648)

Try out this interactivity: Recognising land features (int-3144)

2.5 What landforms are found in the Pacific?

Access this subtopic at **www.jacplus.com.au**

 Deepen your understanding of this topic with related case studies and questions.
 ⊘ Pacific nations

2.6 What processes shape landscapes?

2.6.1 Are all processes natural?

There are processes at work that continuously sculpt and change the landscape. In the future, the Earth's surface will look very different from the way it looks today.

There are a variety of natural processes that shape and reshape not only the surface of the Earth, but also what lies beneath it. Natural processes include uplift, such as that caused by tectonic activity, erosion, deposition and weathering. People change the landscape when they clear land for agriculture or build cities and road networks. Sometimes they alter the course of a river or trap its flow behind the walls of a dam.

2.6.2 What role do tectonic forces play?

The Earth's surface, or crust, is split into a number of plates, which fit together like a giant jigsaw puzzle. These plates sit on a layer of semi-molten material in the Earth's **mantle** — the layer of the Earth between the crust and the core. Heat from the Earth's core creates convection currents within the mantle, causing the plates to move. Most of the Earth's great mountain regions were formed as a result of this movement.

When two plates collide, one plate often slides under the other, in a process known as subduction, and it becomes part of the mantle. Other rocks are forced upwards and bent or folded. Large mountain ranges that were formed in this way include the Himalayas in Asia and the Rocky Mountains in North America. You will find more information on how mountains are formed in subtopic 5.5.

FIGURE 1 Over millions of years, the Colorado River has cut deep channels to form the Grand Canyon.

2.6.3 How is the landscape worn away?

Erosion is the wearing away of the Earth's surface by natural elements such as wind, water, ice and human activity. The landscape is further eroded when agents such as wind, water and ice **transport** these materials to new locations. Eventually, transported material is deposited in a new location. Over time, this material can build up and new landforms result. The Grand Canyon in Colorado in the United States (figure 1) is an example of these elements at work. These processes work more quickly on softer rocks.

Human activity also contributes to erosion. Deforestation, agriculture, urban sprawl, logging and road construction all alter the natural balance and increase erosion by as much as 40 per cent in some areas. Vegetation not only provides valuable habitat for native animals but is also vital for binding the soil together. Once vegetation is removed, it is more easily broken down and removed by wind and water. When topsoil (see figure 3 in subtopic 2.8) is removed, plants are unable to obtain the nutrients they need for growth. Sometimes wide, deep channels, known as gullies, form (figure 2).

FIGURE 2 Note the scale of this gully compared to the people.

FIGURE 3 After tectonic forces cause a section of the Earth to be raised (uplifted), other processes take over and resculpt the landscape.

1 Weathering is the breakdown of rocks due to the action of rainwater, temperature change and biological action. The material is not transported (removed).

It can be physical, chemical or biological.

2 Erosion is the process whereby soil and rocks are worn away and moved to a new location by agents such as wind, water or ice.

3 Transportation is the process that moves eroded material to a new location — examples include soil carried by the wind, sediment or pebbles in a stream.

4 Deposition — materials moved by wind and water eventually come to a halt. Over time new landforms are built. Sand dunes and beaches are common landforms associated with deposition.

Physical — occurs where water is continually freezing and thawing. The water penetrates cracks and holes in the rocks. As water freezes it expands, making the cracks larger. Over time the rock breaks apart.

Chemical — some rocks, such as limestone, contain chemicals that react with water, causing the limestone to dissolve.

Biological — living organisms such as algae produce chemicals that break down rocks. They can also be forced apart by plant roots.

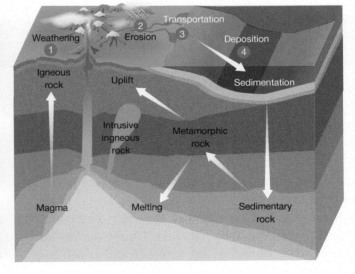

2.6 Activities

To answer questions online and to receive **immediate feedback** and **sample responses** for every question, go to your learnON title at www.jacplus.com.au. *Note:* Question numbers may vary slightly.

Remember

1. In your own words, define the natural processes at work shaping the Earth.

Explain

2. Explain how and why human activity might contribute to weathering and erosion.
3. Using terms such as *uplift*, *erosion*, *deposition*, *weathering* and *transportation*, explain the ***interconnection*** between physical processes and the ***environment***.

Discover

4. Use the internet to discover a landform that you find interesting. Copy and paste this image into a Word document. Annotate your image with information about its location and how it was formed. Add additional annotations to describe how your landform might have ***changed*** over time. Detail how it might have looked in the past and how it might look in the future. Think carefully about the ***scale*** of this ***change***.

Predict

5. Study the ***environment*** around your home or school and find a ***place*** where there is evidence of erosion. Make a sketch and label the features of the landscape. Highlight areas where erosion is evident and add annotations to explain what you think might have caused this ***change***, and in particular, the ***scale*** of this ***change***. Estimate the proportion of this ***environment*** that has been affected. What proportion do you think is the result of human activity? Compare your estimate with the figure given in this subtopic.

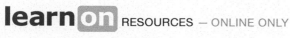

learn on RESOURCES — ONLINE ONLY

Try out this interactivity: Break down! (int-3101)

my**World**Atlas — Deepen your understanding of this topic with related case studies and questions.
❍ **Active Earth**

2.7 What is hidden underground?

2.7.1 What is karst?

Apart from rivers and streams that flow across the surface of the Earth, vast networks of rivers also exist under the ground. The result is a network of caves and channels that carve a very different landscape, known as karst.

Karst is a landscape formed by water dissolving bedrock (solid rock beneath soil) over hundreds of thousands of years (see figure 1). On the surface of the Earth, sinkholes (holes in the Earth's surface), vertical shafts (tunnels), and fissures (cracks) will be evident. Rivers and streams may seem to simply disappear, but underground there are intricate drainage networks, complete with caves, rivers, **stalactites** and **stalagmites** (see figure 2).

Karst topography makes up about 10 per cent of the Earth's surface; however, a quarter of the world's population depends on karst environments to meet its water needs.

2.7.2 How are karst landscapes formed?

Water becomes slightly acidic when it comes into contact with carbon dioxide in the atmosphere (as it does when raindrops form) or when it filters through organic matter in the soil and percolates into the ground. Acidic water is able to dissolve **soluble** bedrock, such as limestone and dolomite. This creates cracks or fissures, allowing more water to penetrate the rocks. When the water reaches a layer of non-dissolving rocks, it begins to erode sideways, forming an underground river or stream. As the process continues, the water creates hollows, eventually creating a cave. Some karst landscapes contain aquifers that are capable of providing large amounts of water.

2.7.3 Where in the world do you find karst landscapes?

Karst landscapes are found all over the world, as shown in figure 3, in locations where mildly acidic water is able to dissolve soluble bedrock such as limestone and dolomite.

In tropical regions, where rainfall is very high, karst mountains sometimes develop. This is because the high rainfall levels wear away the soluble rock much faster than rock is worn away in karst areas with lower rainfall. Examples of tropical karst mountains include the peaks of Ha Long Bay in Vietnam and the Guilin Mountains in China.

FIGURE 1 Formation of a karst landscape

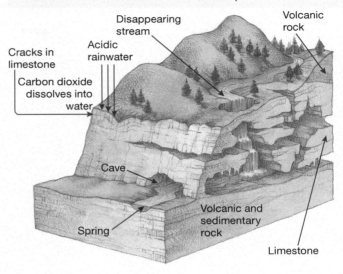

FIGURE 2 Caves in Guilin, Guangxi Province, southern China

The Earth's largest arid limestone karst cave system is located on Australia's Nullarbor Plain, covering 270 000 square kilometres. It extends 2000 kilometres from the Eyre Peninsula in South Australia to Norseman in the Goldfields–Esperance region of Western Australia, and from the Bunda Cliffs on the Great Australian Bight in the south to the Victoria Desert in the north. The extensive cave system provides a unique habitat for a variety of native flora and fauna. Within the caves are fossils that can reveal much about our distant past, along with important Indigenous heritage sites.

FIGURE 3 Karst regions of the world

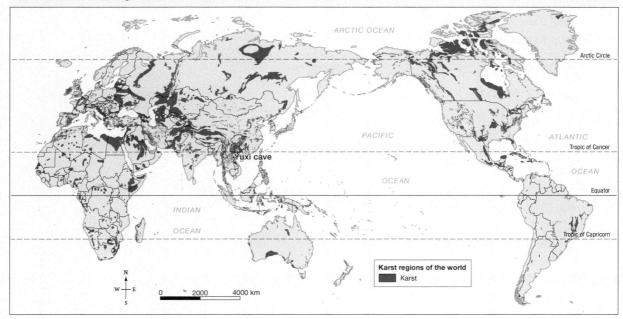

Source: World Map of Carbonate Rock Outcrops v3.0.

2.7 Activities

To answer questions online and to receive **immediate feedback** and **sample responses** for every question, go to your learnON title at www.jacplus.com.au. *Note:* Question numbers may vary slightly.

Remember

1. In your own words, explain how a karst landscape is formed.

Explain

2. Describe the global distribution of karst landscapes.
3. Do you think we should preserve karst landscapes? Give reasons for your answer.

Discover

4. Examples of karst landscapes in Australia include the Buchan, Naracoorte, Jenolan, Labertouche, Princess Margaret Rose, Judbarra and Abercrombie caves. Working with a partner, investigate one of these *environments* and prepare an annotated visual display. Show its location on a map, and include the *scale*, features, land use and any concerns or threats to the *environment*. Include information on what is being done to ensure the *sustainable* management of the *place*. Share your findings with the rest of the class.

Think

5. The largest limestone arid karst system is found on the Nullarbor Plain, Australia.
   (a) The Nullarbor Plain is an example of a desert landscape; suggest how an *environment* formed by water can occur in this location.
   (b) This cave *environment* is a popular destination for caving groups. Use the internet to investigate this *environment* and why people are attracted to it. Compare this *environment* to the one you studied in question 4 above. Pay particular attention to the *scale* and *change* that has occurred in each *place*. Is one more fragile than the other? Explain. Suggest strategies for the *sustainable* management of karst in the Nullarbor.
   (c) Describe how you think this landscape would be different if it were located in Australia's tropical north.

learnon RESOURCES — ONLINE ONLY

Try out this interactivity: Underground wonders (int-3103)

2.8 What's beneath our feet?

2.8.1 What is soil?

We rarely give much thought to the soil beneath our feet. But soil is the basis of all life on the Earth. It provides the nutrients needed for growing plants, which provide food for animals. Without soil, people could not grow crops or raise livestock. Without soil, nothing could survive.

Soil is a thin layer of material on the surface of the Earth. In it, plants can grow. In some parts of the world it is metres deep, but in Australia it is a thin layer of 15 to 20 centimetres. The composition of soil is shown in figure 1 and the factors that influence soil formation are shown in figure 2 in section 2.8.2.

FIGURE 1 While the composition of soil varies widely across the Earth, an average soil will have these characteristics.

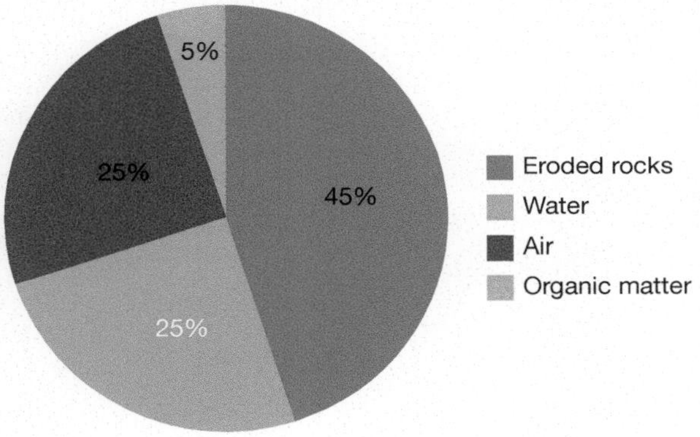
5%
25%
45%
25%
Eroded rocks
Water
Air
Organic matter

Australia generally has poor soils when compared with those found on other continents such as North America and Europe. Australian soils are generally low in nutrients and, in some areas, especially arid zones, they have a high salt content. Patches of good soil, though, are scattered throughout the continent. For example, there is:

- volcanic soil on the Darling Downs in Queensland and at Orange in New South Wales
- alluvial soil in river valleys such as around the Margaret River in Western Australia and the Clarence River in New South Wales.

In many parts of Australia, it takes more than 1000 years for natural processes to produce three centimetres of soil.

2.8.2 How is soil formed?

Factors that influence soil formation are shown in figure 2.

2.8.3 What is a soil profile?

Soil forms in layers called horizons (see figure 3). A soil profile is a side view or cross-section of these different layers or horizons.

FIGURE 2 Influences on soil formation

Climate affects the rate of weathering of soil. In high rainfall areas, soil develops more rapidly, but excess moisture also washes out or leaches nutrients. In rainforests, for example, the rich supply of humus from decaying plant matter produces lush vegetation. However, high rainfall means that without this constant supply of humus, soil fertility is quickly lost. In arid regions, where evaporation is high, soils often contain too much salt to support plant growth. Weather also plays a role; a climate with a freezing and thawing cycle will speed up the breakdown of rocks. In warm climates, the activity of soil organisms is high, and chemical processes also happen more quickly.

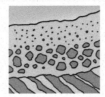

Surface rocks and bedrock are broken down through weathering and erosion. The type of soil that forms depends on the parent material and the minerals it contains. A coarse, sandy soil will develop from sandstone. Bedrock that is mainly granite produces a sandy loam, while shale turns into heavy clay soil.

Soil

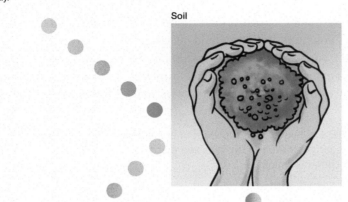

Plants and animals

Decaying plant and animal matter on the soil's surface is broken down by microorganisms into material that is incorporated into the soil, making it nutrient rich.

Topography

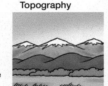

Surface features such as hills, valleys and rivers influence soil development. Soil is generally deeper on the top and at the base of a hill than on its slopes. Floodplains next to river valleys are often nutrientrichdue to sediment being deposited as floodwaters recede.

Time

These processes take place over long periods of time. Soils undergo many changes with the passage of time.

FIGURE 3 A typical soil profile

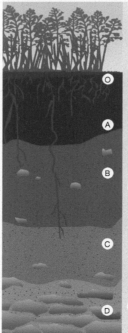

Horizon O (organic matter): Thin layer of decomposing matter, humus and material that has not started to decompose, such as leaf litter

Horizon A (topsoil): The upper layer of soil, nearest the surface. It is rich in nutrients to support plant growth and usually dark in colour. Most plant roots and soil organisms are found in this horizon. In areas of high rainfall, such as tropical rainforests, minerals are leached out of this layer. A constant supply of decomposing organic matter is needed to maintain soil fertility.

Horizon B (subsoil): Plant litter is not present in horizon B so little humus is present. Nutrients leached from horizon A accumulate in this layer, which is lighter in colour.

Horizon C (parent material): Weathered rock that has not broken down enough to be soil. Nutrients leached from horizon A are also found in this layer.

D: Underlying layer of solid rock

2.8 Activities

To answer questions online and to receive **immediate feedback** and **sample responses** for every question, go to your learnON title at www.jacplus.com.au. *Note:* Question numbers may vary slightly.

Remember

1. What is soil?
2. Why is soil important?

Explain

3. In your own words, explain how soil is formed and why it is not uniform across the surface of the Earth.

Discover

4. Use the internet to investigate soils found in desert and rainforest *environments*. Construct a soil profile for each *place* and highlight the differences between them. Find out if the percentages shown in figure 1 are different in each *place*, and add this information to your soil profiles.
5. Use the **Animated soil formation** weblink in the Resources tab. Describe the main steps in the formation of soil.

Predict

6. (a) Dig a hole outside where the soil has not been disturbed too much. Dig until you find small pieces of weathering rock. Measure the depth of your hole. How does this compare with the depth of soils in Australia and overseas?
   (b) Find two pieces of rock that show signs of weathering. Check the hardness of these rocks; the harder the rocks, the more difficult it will be to obtain a sample. Rub them together over a piece of paper. Were you able to collect a spoonful of grains in a reasonable amount of time? If so, how long did it take to rub off a spoonful of particles?
   (c) The rate of soil formation is estimated at less than 0.05 millimetres a year in eastern Australia. How long would it take to develop one centimetre of soil? How long would it take to form enough soil to replace what was in the hole you dug earlier?
   (d) Write a paragraph explaining what this exercise tells you about soil formation and the need to use soil in a *sustainable* manner.

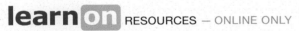 RESOURCES – ONLINE ONLY

 Explore more with this weblink: Animated soil formation

2.9 SkillBuilder: Using positional language

WHAT IS POSITIONAL LANGUAGE?

Positional language is the use of compass points to locate places and provide directions between places. North, north-east, east, south-east, south, south-west, west, and north-west are shown on an 8-point compass. We can use positional language to describe the location of one feature in relation to another.

Go online to access:

- a clear step-by-step explanation to help you master the skill
- a model of what you are aiming for
- a checklist of key aspects of the skill
- a series of questions to help you apply the skill and to check your understanding.

FIGURE 1 An eight-point compass

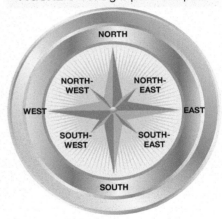

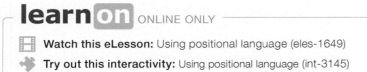
2.10 How do different cultures view landscapes?

2.10.1 The Australian context

Landscapes are the product of processes that have operated for millions of years. While all humans have come to realise the importance of the landscape and the role it plays in our lives, indigenous groups were the first to recognise that it is important to work with nature rather than always seek to change and exploit it.

Aboriginal peoples and Torres Strait Islander peoples are recognised as the first Australians. Evidence of their presence in Australia is found across the continent

FIGURE 1 Rock art depicting a cloud or rain spirit, found in Western Australia's Kimberley region

FIGURE 2 Map of Kakadu National Park

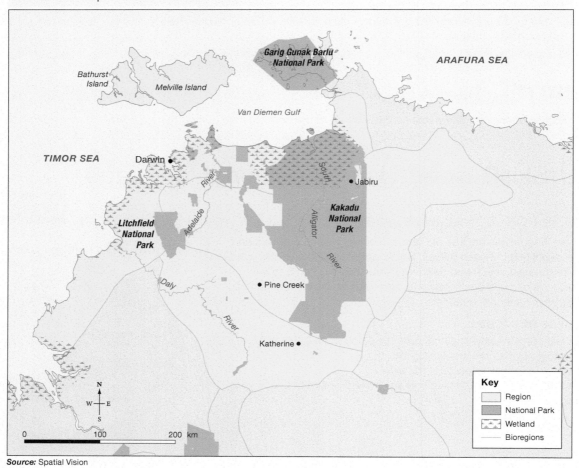

Source: Spatial Vision

in their rock art (as shown in figure 1), in **archaeological** records, and through their cultural heritage passed down through generations. As **hunter–gatherers** they relied on the plants, animals and the environment for their survival, and so have an understanding of the complex nature of Australia's varying landscape.

Europeans, on the other hand, arrived in 1788 and occupied the area. They had a very different view of the landscape, based on ideas they brought with them from Britain. They sought to change the landscape and adapt it to meet their needs. They established permanent settlements and depended on agriculture to provide for their needs.

The perspective of Indigenous Australian peoples is one of being part of the landscape, while the European perspective is based on the idea of land ownership. Use the **One Night the Moon** weblink in the Resources tab to find out more about the different ways in which the landscape is viewed.

2.10.2 Kakadu — Australia's first World Heritage Area

Kakadu National Park, as seen in figure 2, covers an area of approximately 20 000 square kilometres of the Northern Territory — an area roughly a third the size of Tasmania. It stretches 200 kilometres from north to south, and spans 100 kilometres from east to west. Within the boundaries of the park are vast uranium deposits. Kakadu is unique in that it is recognised for both its natural beauty and its cultural value.

FIGURE 3 Why Kakadu is valued

FIGURE 4 Jim Jim Falls at Kakadu is a popular tourist destination.

2.10.3 Kakadu and its resources

Kakadu is rich with the historical records and ancestry of the first Australians. In addition, it supports a treasure trove of native plant and animal species and provides a temporary home to a large number of migratory birds. More than 170 000 tourists visit Kakadu annually, attracted by its vast wetlands and scenery, including steep gorges, Aboriginal rock art, waterfalls such as Jim Jim Falls (see figure 4), and lookouts.

Kakadu also has vast deposits of uranium ore, which some regard as a valuable export for Australia. Opponents of uranium mining are concerned about the possibility that Australia's uranium could be processed and used to make nuclear weapons. Others fear the effects of mining on the environment and the potential for a devastating pollution event.

The Ranger uranium mine has been operating for 30 years and lies within the boundaries of Kakadu National Park. Three kilometres downstream from the mine, the Mirrar people (a local Aboriginal community) swim and fish. Since the mine opened, there have been more than 200 leaks and spills, and the mine has generated some 30 million tonnes of liquid radioactive waste (see figure 5).

FIGURE 5 Timeline of major breaches at the Ranger uranium mine since 2002

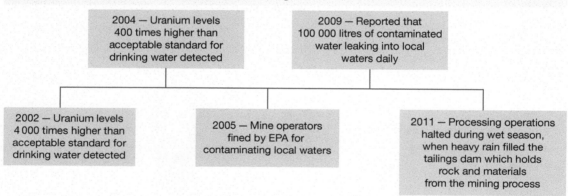

2004 — Uranium levels 400 times higher than acceptable standard for drinking water detected

2009 — Reported that 100 000 litres of contaminated water leaking into local waters daily

2002 — Uranium levels 4 000 times higher than acceptable standard for drinking water detected

2005 — Mine operators fined by EPA for contaminating local waters

2011 — Processing operations halted during wet season, when heavy rain filled the tailings dam which holds rock and materials from the mining process

2.10 Activities

To answer questions online and to receive **immediate feedback** and **sample responses** for every question, go to your learnON title at www.jacplus.com.au. *Note:* Question numbers may vary slightly.

Remember

1. Where is Kakadu National Park and why is it important?
2. Copy the table below into your workbook and use it to compile a list of differences in the way the Australian landscape was viewed by Aboriginal and Torres Strait Islander peoples and non-Aboriginal and Torres Strait Islander Australians. The first one has been done for you.

Aboriginal and Torres Strait Islander peoples views	Non-Aboriginal and Torres Strait Islander views
The land is communally owned.	Individuals own the land.

Explain

3. (a) Where are the more densely populated regions of Australia? *Hint:* Find a map in your atlas that shows population distribution.
   (b) Why would it be more difficult for Aboriginal and Torres Strait Islander communities in these areas to maintain their traditional lifestyle and culture?
4. Describe the *interconnection* that Aboriginal and Torres Strait Islander peoples have with the landscape. What evidence of this *interconnection* is found in this subtopic and in the **One Night the Moon** weblink in the Resources tab?
5. What is uranium used for and why is it considered a valuable resource?
   (a) What risks does uranium mining pose in the Kakadu region?

Think

6. Think about your personal values and beliefs and analyse how they might be similar or different to those reflected in figure 3.

Predict

7. (a) Suggest three possible impacts on the landscape if a new uranium mine was opened in the Kakadu region.
   (b) Do you think *changes* would have a large-*scale* or a small-*scale* impact? Explain.
8. Why do you think the Australian government allows uranium mining in such an important region of Australia?
9. Predict what pressures decision-makers in Australia might face in future when balancing the needs of the different groups who have an interest in Kakadu's resources.
10. Write a letter to the editor of a newspaper outlining your views on uranium mining in *environmentally* sensitive areas. Explain whether you consider this type of activity is a *sustainable* use of the landscape.

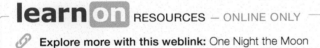

learn**on** RESOURCES — ONLINE ONLY

🔗 **Explore more with this weblink:** One Night the Moon

2.11 How are landscapes preserved and managed?

2.11.1 The World Heritage Convention

Worldwide, people recognise the value of landscapes and the need to protect their natural beauty and cultural heritage, and to manage their resources sustainably. Landscapes are easily damaged or destroyed but are difficult to recreate and repair. The key is to ensure that they are carefully managed so that the landscapes we value today are still present in the future.

From the middle of the twentieth century, there was growing concern about the need to protect areas of both cultural and natural significance (see figure 1).

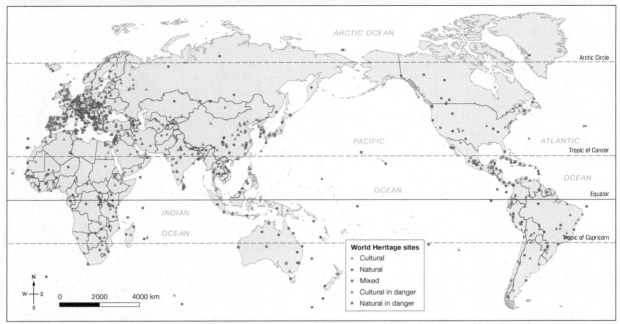

2.11.2 The Artesian Range

The Artesian Range is a unique part of the Australian landscape. It has been described as a lost world, a modern-day Noah's Ark, our last opportunity to protect and preserve a part of the Australian mainland that has had little contact with modern civilisation. Within its hidden valleys and canyons lies a diverse range of flora and fauna. The rich tropical rainforests and woodlands provide vital habitats for some of Australia's most endangered wildlife.

The Artesian Range covers 1800 square kilometres (see figure 2). It is largely inaccessible; the only way in is by helicopter or boat. It is a maze of hidden valleys and canyons, rocky ranges and plateaus, towering **escarpments**, wide valleys and deep gorges (see figure 3). Its sandstone ranges were formed as a result of moving tectonic plate activity. These rock formations date back some 1.8 million years.

FIGURE 2 The Artesian Range covers 1800 square kilometres of the Kimberley region.

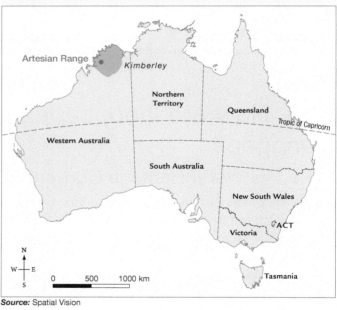

Although it is difficult for humans to reach the area, exotic species such as donkeys, horses, pigs and cats have gradually invaded the Kimberley. And while fire is a natural part of the landscape, changing fire patterns and the increasing number of late-season wildfires are also a threat to the Artesian Range. Australian Wildlife Conservancy (AWC), an independent non-profit organisation funded by donations, has now secured the land, and manages it for conservation. AWC undertakes fire management, feral animal control, and biological surveys and monitoring, protecting the full length of the Artesian Range.

FIGURE 3 The Artesian Range is a rugged and largely inaccessible landscape, renowned for its natural beauty and unique wildlife.

Source: AWC/Wayne Lawler

2.11 Activities

To answer questions online and to receive **immediate feedback** and **sample responses** for every question, go to your learnON title at www.jacplus.com.au. *Note:* Question numbers may vary slightly.

Remember

1. Why is it important to protect sites that have cultural or natural significance?
2. Describe the location of the Artesian Range and why it is unique.

Explain

3. Suggest why the Artesian Range has been largely inaccessible to people.
4. Explain how exotic species such as cats, foxes and camels have been able to become established in the Artesian Range when it is difficult for people to enter the region.

Discover

5. Use the **World Heritage list** weblink in the Resources tab and select a site in one of the countries listed on the map. Prepare a visual presentation of one of the sites listed, outlining its importance and how it is protected.

Predict

6. In small groups, investigate an invasive species and describe the ways in which it has *changed* the *environment*. Is this *change* occurring on a small or large *scale*? Explain. Suggest a strategy that the Australian Wildlife Conservancy could employ to eradicate invasive species from this *environment*.

Think

7. (a) Explain what you understand by the terms *cultural significance* and *natural significance*.
   (b) Is it possible for **places** to have both cultural and natural significance? Draw up a table like the one below. With the aid of a partner, add as many **places** as you can to the list. Try to have a balance of Australian and international examples. Compare your list with that of another pair of students.

Cultural significance	Natural significance	Cultural and natural significance

   (c) Which column has the most entries? Suggest a reason for the pattern you observe.
   (d) Select one **place** from column 3. Find a picture of this **place** and copy and paste it into a Word document. Add annotations to explain the major features of your chosen **place** and why it is of cultural and natural significance.
8. Evaluate the ways in which the community demonstrates the value it places on cultural diversity and why this is important to the community.

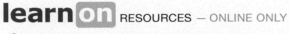 RESOURCES — ONLINE ONLY

⌗ **Explore more with this weblink:** World Heritage list

 **Deepen your understanding of this topic with related case studies and questions.**
   ❯ **World Heritage sites**

2.12 Review

2.12.1 Review
The Review section contains a range of different questions and activities to help you revise and recall what you have learned, especially prior to a topic test.

2.12.2 Reflect
The Reflect section provides you with an opportunity to apply and extend your learning.

Access this subtopic at **www.jacplus.com.au**

TOPIC 3
Landscapes formed by water

3.1 Overview

Numerous **videos** and **interactivities** are embedded just where you need them, at the point of learning, in your learnON title at www.jacplus.com.au. They will help you to learn the content and concepts covered in this topic.

3.1.1 Introduction

Water is one of the most powerful agents in creating landscapes. If you have ever been caught outside in a heavy downpour, walked through a fast-flowing creek, or been dumped in the surf, then you have felt and seen the energy of flowing water. It can knock you off your feet, move buildings and carve huge holes in the Earth's surface. Landscapes created by water are found everywhere.

Pacific Ocean waves pounding against lava cliffs

learn **on** RESORCES — ONLINE ONLY

🔗 **Explore more with this weblink:** Iguazu Falls

3.2 Which landscapes are formed by water?

3.2.1 How does water change landscape features?

A torrent of gushing water can shift rocks, remove topsoil or shape river valleys. Gentle rain can change the chemical structure of any surface material, making it more likely that soil will be transported by the next heavy shower. In cold climates, frozen water in glaciers works like a slow-moving bulldozer to erode land and create unique landscape features. Once fresh water has made its way to the ocean, the power of waves creates coastal landscape features.

As you saw in topic 2, landscapes are predominantly changed or created by two processes: erosion and deposition. Water is one of the most powerful agents of change, causing erosion and deposition, and thereby breaking up, moving and repositioning material across the Earth's surface. In figure 2 you can see the power of water as it rushes over a rockface and carves pools in its hard surface. You may have seen pools of a similar shape carved by waves in rocky coastal landforms.

FIGURE 1 The Twelve Apostles in Port Campbell National Park, Victoria. How might the potential for erosion change along this coast if the waves were larger and it was high tide?

Stacks show where the coastline used to be.

Current coastline

Collapsed stacks

Limestone cliffs

As water makes contact with landscapes, it can change the shape and size of its features or landforms (figures 2 and 3). The coastal landscape that you see today is not the same as it was hundreds or thousands of years ago. Figure 1 is a photo of the Twelve Apostles, located on the coast of south-western Victoria. The name suggests that there may once have been twelve pillars of rock, or stacks, visible along this stretch of coastline. In the foreground you can see the remnants of two quite recently collapsed stacks. Even these stacks were once joined to the cliffs as part of the mainland. This highly erodible coastline has been constantly altered by many years of rainfall and wave action on the soft limestone cliffs. The resulting coastline has seen much creation and destruction of stacks over time.

FIGURE 2 How is the flow of water changing this landscape?

FIGURE 3 Water constantly moves over and through the Earth and through the air.

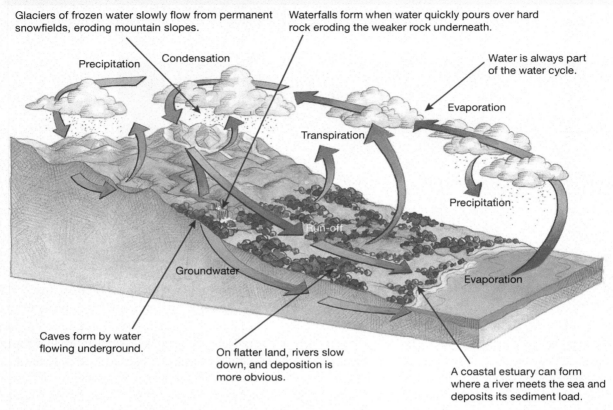

Glaciers of frozen water slowly flow from permanent snowfields, eroding mountain slopes.

Waterfalls form when water quickly pours over hard rock eroding the weaker rock underneath.

Water is always part of the water cycle.

Precipitation

Condensation

Evaporation

Transpiration

Precipitation

Run-off

Groundwater

Evaporation

Caves form by water flowing underground.

On flatter land, rivers slow down, and deposition is more obvious.

A coastal estuary can form where a river meets the sea and deposits its sediment load.

3.2 Activities

To answer questions online and to receive **immediate feedback** and **sample responses** for every question, go to your learnON title at www.jacplus.com.au. *Note:* Question numbers may vary slightly.

Remember

1. Landscapes are in a state of continual *change*.
   (a) Which two natural processes powered by water are most responsible for continually *changing* landscapes?
   (b) How are these two processes linked?

Explain

2. Where would figures 1 and 2 be *placed* on the landscape depicted in figure 3? Explain.
3. Explain how the water cycle and the formation of landscapes are *interconnected*.

Discover

4. Use your research skills to create a list of world water facts on the following:
   (a) the biggest glacier
   (b) the longest river
   (c) the biggest wave
   (d) the highest waterfall
   (e) the widest river
   (f) the biggest ocean
   (g) a world water fact of your choice.
   Show on a map where each is located.

5. Many landscapes *change* rapidly; for example, the Twelve Apostles. With a partner or group, discuss another example of a landscape that has been shaped by the power of water. Do you think the *changes* to the landscape have been positive or negative? To what extent should people try to stop the *changes* caused by water?

6. Water can be considered one of the most important architects of desert landscape features. After looking at the images on this page, try to explain how you think water can change the landscapes of arid or desert environments.

7. Identify three possible ways that people can *change* the flow of water, either across the surface of the Earth or along the coast. Predict how you believe this may alter landscape features. Examples may include the use of river water for irrigation or the construction of a marina.

3.3 What is coastal erosion?

3.3.1 How are coastal landscapes eroded?

The coast is the zone or border between land and ocean. It is in this collision zone that the movement of sea water and the impact of the ocean on the land together create coastal landscapes. Coastal landscapes have landforms that are common to coastlines in different places around the world because they are built up or worn away in similar ways.

Powerful ocean waves crash onto rocky coastlines, wearing away the cliff base between the high and low tide marks. In figure 1 in subtopic 3.2, showing the Twelve Apostles, the most powerful wave impact on this coast would be to the seaward side of the stacks or rock pillars. Wave impact has progressively knocked over some stacks, and the rock that they were made from has eroded into sand, which now forms part of the beach at the cliff base.

Coastal erosion is mostly caused by waves moving sand and other material and energy to and from the beach. For example, you may have seen sand being churned up by the waves when you have been at the beach. The waves that wash on to the beach are called **swash**. Waves returning down the beach into the sea are called **backwash**. A powerful ocean swell created by offshore storms forms **destructive waves**. These are the waves that produce good surf, and dump you while you are swimming. As these waves rush back into the sea, the force of the water and sand they carry can make it difficult for you to get back onto the beach. Destructive waves often have a powerful backwash that can carry beach sand offshore. Figure 1 shows backwash filled with sand that is being moved from the beach.

FIGURE 1 Backwash filled with sand

Direction of water movement

Backwash filled with sand

However, not all aspects of coastal landscapes are solely created by the power of water. Other physical processes can also greatly affect the coastal landscape; for example, the tectonic force of earthquakes and volcanoes; changing sea levels; and human activities such as building roads, ports and houses, and damming rivers.

Which coastal landscape features are created by erosion?

Features such as cliffs, headlands, bays, caves and stacks are all landforms found along an eroding coastline (figure 2). These features are formed by wave action and rainfall, which attack the cliffs and find points of weakness that are then eroded. Water running off a cliff face can carry eroded material into the sea below. When waves hit the cliff face, they undercut the base of the cliff to form a notch. As the notch increases in size, the undercut section of the cliff becomes unstable and falls into the sea.

Destructive waves can also alter a sandy coastline. They can remove sand from a beach, destroy the vegetation on dunes, and remove management features designed to protect landscape features.

FIGURE 2 Coastal landforms created by erosion

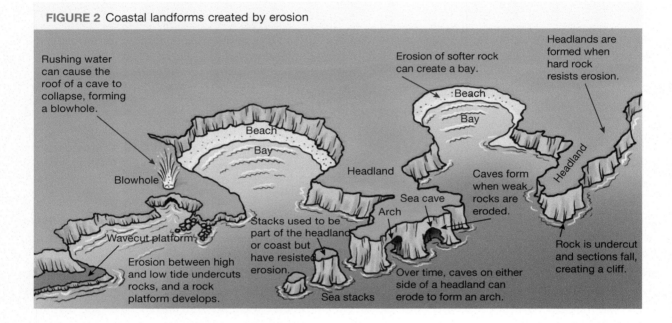

3.3 Activities

To answer questions online and to receive **immediate feedback** and **sample responses** for every question, go to your learnON title at www.jacplus.com.au. *Note:* Question numbers may vary slightly.

Remember

1. What are three physical processes that have influenced the creation of coastal landforms?
2. What are three human activities that have influenced the creation of coastal landforms?

Explain

3. Create an annotated diagram that explains the difference between swash and backwash.
4. Place the following landforms in the order in which they would be created:
   (a) arch, cave, headland, stack
   (b) blowhole, cave, cliff.

Discover

5. Watch the **Stack formation** weblink and/or the **Cliffed Coast** weblink in the Resources tab. Take note of the process of erosion of a cliff face.
6. In a small group, create your own claymation or stop-motion movie, Prezi, or animated PowerPoint to show the *changes* that happen to a cliffed coast eroding to form a notch, cave, arch and stack.
7. Find an image of a sandy coastline that has recently been affected by destructive waves. Explain the process that has occurred. Use the terms *swash* and *backwash* in your explanation.

Predict

8. Most Australians live within an hour's drive of the coast, and many people either spend regular holidays on the coast or move to the coast in their retirement, for a 'sea change'. How might the continually *changing* coastal landscape (as seen in figure 2) affect coastal housing and popular holiday *places*? Brainstorm this with a small group.
9. Identify, using a sketch map, how several of the *changes* identified in question 8 might affect the coastal landscape of your favourite beach.
10. Do you think people will still feel the same way about a coastal landscape such as the Twelve Apostles when only two or three are still standing? How might the *changing* landscape affect the value or pleasure people get from visiting this *place*? Write a short paragraph to comment.

Think

11. Rising sea levels, whether they are a naturally occurring process or have resulted from human activity, will affect coastal landscapes. Use a diagram, with annotations, to explain how rising sea levels could *change* two of the landforms illustrated in figure 2.

 Try out this interactivity: Coastal sculpture (int-3128)

 Explore more with these weblinks: Stack formation, Cliffed Coast

 Deepen your understanding of this topic with related case studies and questions.
❯ **Coastal processes**

3.4 Which coastal landforms are created by deposition?

3.4.1 How are depositional coastal landforms formed?

Not all waves are destructive. Some waves gently lap the shoreline. Smaller, gentler waves that carry less energy than destructive waves are known as **constructive waves**. The movement of these waves towards the land is more likely to push material such as sand and shells and deposit them on the beach, building new coastal features.

A beach is a good example of a depositional coastal landform (figure 1). Sand has been deposited and built up over a period of time. Constructive waves build coastal landscape features by repositioning wave-born materials to also create spits, sand dunes and lagoons.

The coastal features created by deposition can be created only when material is brought onshore by the swash of constructive waves. The construction material is in the form of sand, shells, coral and pebbles.

The source of the construction material may come from eroding cliffs, from an offshore source, or from rivers which, when they enter the sea, dump any material they were transporting.

This construction material is then shaped by prevailing winds. Figure 2 illustrates the cross-section of a beach formed when there is plenty of sand being pushed onshore by the swash. This construction material is dried by the sun and blown inland to create dunes.

Beach material can also be shifted by waves, which get their energy from the wind. The wind influences or directs the angle that waves move towards the coast. Waves come from the direction of the **prevailing wind**. This means that waves often move towards the shore at an angle, and their swash pushes any material they are carrying onto the beach at an angle. As the backwash of the wave returns to the sea, its path takes the shortest possible route down the beach towards the water. This action is known as longshore drift, and it is shown in figure 3. Longshore drift moves material along the beach in a zigzag pattern that follows the direction of the prevailing wind. Longshore drift moves sand along the beach and creates spits and bars. If the prevailing wind changes direction, then so does the direction of longshore drift.

FIGURE 1 Depositional landforms: coastal landforms created by deposition

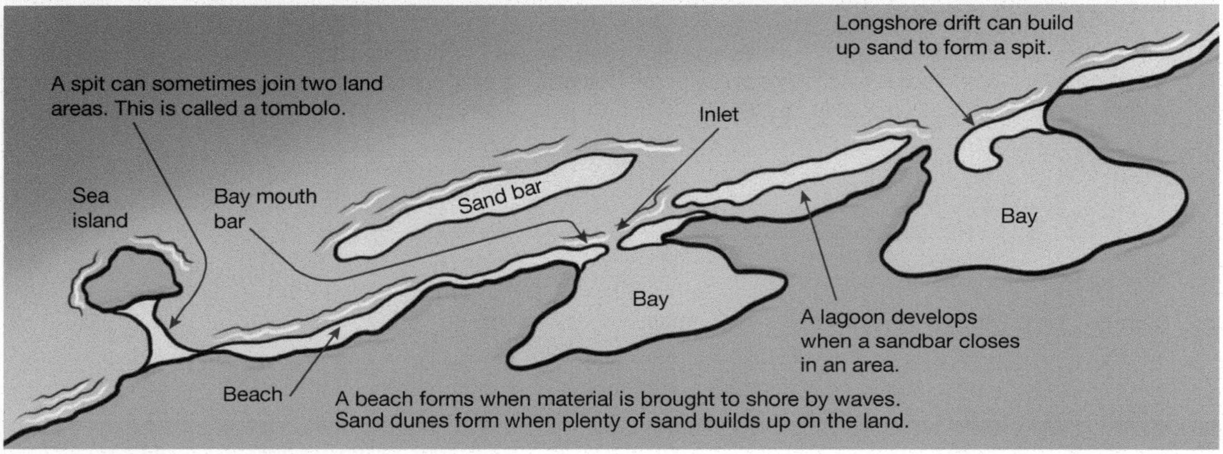

Longshore drift can build up sand to form a spit.

A spit can sometimes join two land areas. This is called a tombolo.

Inlet

Sea island

Bay mouth bar

Sand bar

Bay

Bay

A lagoon develops when a sandbar closes in an area.

Beach

A beach forms when material is brought to shore by waves. Sand dunes form when plenty of sand builds up on the land.

FIGURE 2 The formation of sand dunes

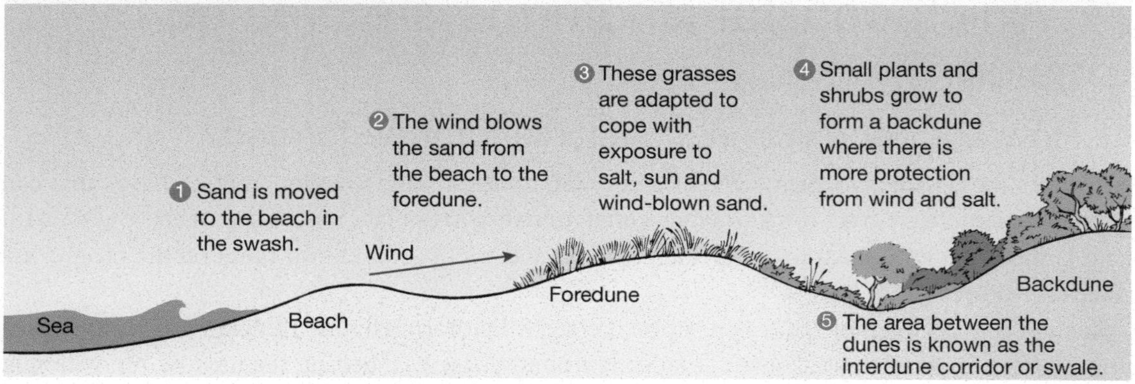

❸ These grasses are adapted to cope with exposure to salt, sun and wind-blown sand.

❹ Small plants and shrubs grow to form a backdune where there is more protection from wind and salt.

❷ The wind blows the sand from the beach to the foredune.

❶ Sand is moved to the beach in the swash.

Wind

Foredune

Backdune

Sea

Beach

❺ The area between the dunes is known as the interdune corridor or swale.

FIGURE 3 The process of longshore drift

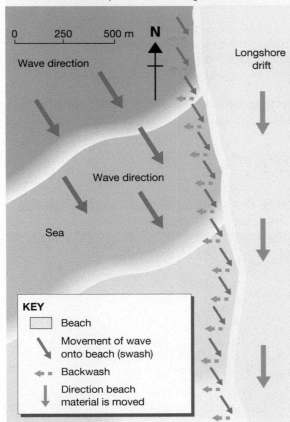

KEY

☐ Beach

↘ Movement of wave onto beach (swash)

◀-- Backwash

↓ Direction beach material is moved

FIGURE 4 The depositional landforms of the coastal landscape at Tutukaka in Northland, New Zealand

CASE STUDY

The Murray mouth, South Australia

The Murray River is Australia's most important river and the world's sixteenth longest river.

When water for home use and irrigation in the Murray–Darling Basin is not balanced by rainfall, the amount of water that reaches the river mouth decreases. This means that the deposition of longshore drift is stronger than the trickle of water reaching the mouth. In 2009, $24 million was allocated by the Federal government to dredge the mouth of the river in an effort to keep it open.

3.4 Activities

To answer questions online and to receive **immediate feedback** and **sample responses** for every question, go to your learnON title at www.jacplus.com.au. *Note:* Question numbers may vary slightly.

Remember

1. Where does the material come from that builds beaches?
2. Either scan and enlarge a copy of figure 4 or draw a sketch of it, then label or annotate it to identify the depositional landform features of this coastal landscape.

Explain

3. The formation of sand dunes cannot happen unless there is plenty of sand in the swash to allow them to grow. Use the information in figure 2 to provide the evidence for you to agree or disagree with this statement.
4. How is weather involved in the formation of sand dune *environments*?

Discover

5. Use your atlas plus the internet to locate and name *places* in Australia that have the following coastal landforms: a spit, a beach with dunes, a bay, a headland (point, cape or promontory) and an estuary. Find four examples of each landform and mark them on a map. You could create a Google map of your results, with links to images of each feature.

3.5 How are coasts changed?

3.5.1 Who uses the coast?

Coasts are very popular places for people to live near and visit. The first Europeans to Australia arrived by sea, and the first areas inhabited by the Europeans were within coastal landscapes. However, the increasing popularity of coastal environments can result in crowding, pollution, a loss of **habitat**, and the building of permanent structures in zones that should be allowed to undergo the natural changes of a coastal landscape.

Aboriginal peoples and Torres Strait Islander peoples travelled from what is now Asia to the north, first settling along the coast. They enjoyed coastal food **resources**, such as fish, shellfish and sea birds;

FIGURE 1 Shell midden in the dunes of the Coorong National Park, South Australia

nearby forest resources for food and shelter; and the moderate coastal climate. Figure 1 shows a coastal landscape feature built by Indigenous peoples. Centuries of regular camps in these dunes resulted in the construction of a midden from the discarded shells of meals of oysters and cockles. To encourage sustainable harvesting of coastal resources, the top layer of the midden showed the next visitors what had been eaten most recently. They would then know to eat something else and not overuse a single food source.

There are many thousands of similar sites in all Australian states and in other countries. They are evidence of the use and habitation of coastal landscapes by people for thousands of years.

Since 1788, the coastline of Australia has been used as:

- the main highway linking Australian cities to other places in Australia
- a place to dispose of liquid and solid waste
- a holiday playground
- a resource for fishing
- a place for mining
- a very popular scenic area.

CASE STUDY

Cape Woolamai

Cape Woolamai, Phillip Island is a well-known surf beach location in south-eastern Victoria. The Cape is pounded by the westerly swells off Bass Strait on its southern coast, and the Eastern Passage, entrance to Western Port Bay, protects its northern coast. Cape Woolamai was created when a build-up of sand joined the coast to a small granite outcrop with a tombolo or narrow sandy isthmus. This sandy environment was stabilised by many centuries of vegetation growth.

Tourists, surfers and beach goers have used this place for many years and their use has affected the landscape quality. Walking and driving on the dune vegetation has resulted in vegetation destruction and a number of blowouts along the dunes. The surf beach access tracks have eroded, with sand being blown inland and depleting the beach of sand. On the more protected northern side, sand surfing has destroyed vegetation. As can be seen in figure 2, a large section of dune vegetation has gone. People on surfboards skimming down the dune have caused this vegetation loss. The popularity of this place has seen the addition of an access road, large car park, boardwalks and protective fencing in an attempt to manage landscape degradation.

FIGURE 2 Human landscape degradation caused by sand surfing at Cape Woolamai Beach, Phillip Island

3.5.2 Why does a coast need to be managed?

The popularity of a coastal location, and the many competing ways people use it, means that it must be looked after to protect it from damage caused by overuse (see figure 3).

FIGURE 3 Why the coast needs to be managed

Coastal landscapes can also be changed by wave action from strong winds. The roads and buildings located on both rocky and sandy coasts are vulnerable to damage from attack by destructive waves.

However, it is possible to manage or protect the natural and built features of coastal landscapes from the physical processes that strive to alter them.

3.5 Activities

To answer questions online and to receive **immediate feedback** and **sample responses** for every question, go to your learnON title at www.jacplus.com.au. *Note:* Question numbers may vary slightly.

Remember

1. Why is it important for people to manage their use of the coastal landscape?
2. List as many examples of coastal use and management as you can identify in figure 3.

Explain

3. How did Indigenous Australian peoples protect the features of the landscape along the coast?
4. Why does the material that a coastal landscape is made from affect the ability of water to shape it? Compare a cliffed coast like the Twelve Apostles to a surf beach with dunes. Refer to subtopics 3.2 and 3.4.
5. Carefully look at figure 2 and identify the evidence of human impact on this *place*.
6. How would the landscape management strategies of building a road, car park, boardwalks and protective fences reduce future landscape degradation of the dunes at Cape Woolamai?

Think

7. Describe the *interconnection* between sand surfing and vegetation loss on the Cape Woolamai dunes.
8. Write a paragraph to explain how attempts to reduce or manage coastal landscape destruction could reduce the recreational experience of visitors.

3.6 What differences exist in coastal landforms between places?

3.6.1 How do coastal landforms differ?

Although coastal landforms can be similar in different parts of the world, they can also be very different. Some differences are climatic and some are geomorphic. Coastal landscapes are created by the interconnections between the sculpting power of the oceans, coastal topography and the material that is available to sculpt.

Limestone stacks, such as the Twelve Apostles in Victoria (figure 1, subtopic 3.2), have been shaped by the power of the Southern Ocean. Similar stacks have been formed by the erosive power of the waters off the coast of Thailand (figure 1) and along the Portuguese and Welsh coasts.

The Gippsland Lakes in south-eastern Victoria are a network of coastal lakes and lagoons fed by six rivers but they are often cut off from the sea by a barrier of silt. The Gippsland Lakes are at the mouth of the Mitchell, Avon, Thompson, Latrobe, Nicholson and Tambo Rivers. When there is little rainfall, the rivers flow slowly and

FIGURE 1 Ko Tapu rock near Phuket, Thailand

deposit sediment in the lakes. This, along with the longshore drifting of the sea current in Bass Strait, creates lakes by moving sediment to seal the lakes with offshore barriers. After heavy rainfall the level of water in the Lakes rises and the barrier breaks, allowing access of fresh water to the sea and salt water into the Lakes. This lake system had an artificial entrance cut by humans in the late 1800s to allow fishing boats into and out of the Gippsland Lakes and to reduce the chance of algal blooms.

In south-eastern Iceland the melting Vatna-jökull glacier (figure 2) flows into the Atlantic Ocean through a glacial lake. This glacier once flowed directly into the sea, but a warming local climate has meant that the glacier's snout is now 1.5 kilometres inland. The melting ice has created the large 18-square-kilometre glacial lake named Jökulsárlón. Since the climate is cold and the sunshine has little heat, the large chunks of ice that fall from the glacier remain as slowly melting icebergs. These icebergs float in the lake until they become small enough to roll down a channel into the sea. During winter the lake freezes and traps the icebergs until the summer thaw. Humans have created a narrow channel to

FIGURE 2 Jökulsárlón Glacier Lagoon, Iceland

link Jökulsárlón with the sea. This channel is designed to reduce the chance of summer floods and to protect the major highway that brings tourists to this beautiful place.

These two coastal lakes have formed in very different places, with different climates, but the geomorphic process of deposition has meant that human intervention has been required to allow their waters to flow into the sea.

3.6 Activities

To answer questions online and to receive **immediate feedback** and **sample responses** for every question, go to your learnON title at www.jacplus.com.au. *Note:* Question numbers may vary slightly.

Remember

1. What material are the Twelve Apostles and Ko Tapu rock both made from?

Explain

2. How has climate *changed* the entrance of the Vatnajökull glacier into the sea?
3. Describe the way that the geological process of deposition has *changed* the Gippsland Lakes and Jökulsárlón.

Discover

4. Use the internet to collect at least six images of limestone stacks from different *places* in the world.
5. Attach these images to a Google map to create a global distribution of limestone landscapes.
6. Describe the similarities and differences between the images.

Predict

7. The Vatnajökull glacier is expected to have melted within 80 years. What might this *place* look like when there is no longer a glacier? Draw a sketch map to explain your answer.
8. Look at a map of the Gippsland Lakes. Predict how they might look if part of the barrier washes away during a huge storm. Draw a sketch map to explain your answer.

3.7 How are coasts managed?

3.7.1 How can a coast be managed?

It is possible to reduce or slow the change to coastal landscapes if we understand the **physical processes** and human activities that cause it. While it is not possible to change the speed and direction of the wind or the number of months each year when destructive waves reach a shoreline, it is possible to redistribute or trap the sand shifted by storm waves or longshore drift. It is also possible to protect coastal houses and roads using barriers to reduce the direct impact of waves.

Some of the most common structures built to protect coastal landscapes and manage change are sea walls and breakwaters, as seen in figure 1, or rock barriers and groynes as seen in figure 2.

FIGURE 1 The coastal village of Pennan, Scotland, is protected by its sea wall and breakwater.

Sea wall

Breakwater

FIGURE 2 A groyne and rock barrier protect a sandy beach in Wales, United Kingdom.

Groyne

Rock wall

Do these management strategies always work?

An integrated strategy like the one designed for Adelaide's beaches has a much better chance of protecting existing coastal landscapes (particularly the beaches) and structures built nearby, because it has taken into account the prevailing wind conditions, as well as the movement of sand. If a structure like the groyne in figure 2 is built on a beach, it will certainly trap sand on the side that interrupts the direct flow of the longshore drift. But this structure will also reduce the flow of sand to beaches further along the coast, on the other side of the groyne. Building a sea wall or breakwater may interrupt the flow of longshore drift and actually silt up the mouth of the harbour it is protecting. A sea wall can deflect the power of waves and increase erosion on an unprotected part of the nearby coast, or reduce the erosion of material from a cliff face that had been replenishing sand on the local

FIGURE 3 Piping sand from north to south along Adelaide's beaches

beaches. Coastal management is quite a tricky issue. Do you manage to protect the existing coastal landscape or do you manage to allow the action of wind and waves to create a naturally evolving landscape?

3.7.2 Adelaide's coast

CASE STUDY

Managing Adelaide's living beaches

The problem: The beautiful sandy beaches closest to Adelaide are under constant threat from erosion. Figure 4 identifies the problem. For the past 7000 years the beaches south of Adelaide have been eroding, and the prevailing winds from the south-west have driven this material northwards. This longshore drift has removed material from the south and relocated it in North Haven, where a **peninsula** has grown and a large dune system has been created. For the past 30 years the beaches in the south have been replenished by adding truckloads of sand. The plan is to find a better way to manage Adelaide's beaches by reducing the cost of moving sand.

The solution: Adelaide's Living Beaches Strategy. Figure 5 illustrates the solution. Although sand will still need to be recycled from north to south, the plan is to use a pipeline instead of trucks to do most of the transportation. The pipeline will extend along the coast and will send sand back to the southern end of the beach. Figure 3 shows sand being discharged at the southern end of the beach. A series of structures such as breakwaters and groynes will be built in several places to trap sand at important locations. Fewer trucks will be used, and it is expected that the cost of beach restoration will be reduced.

FIGURE 4 The movement of sand northwards along the Adelaide Metropolitan coastline

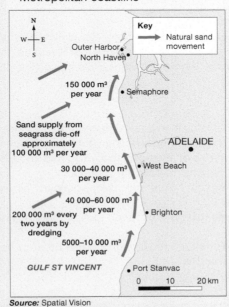

Source: Spatial Vision

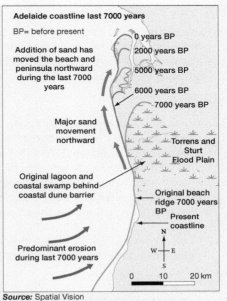

Source: Spatial Vision

FIGURE 5 Adelaide's Living Beaches Strategy

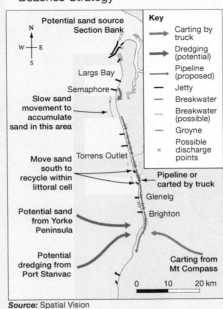

Source: Spatial Vision

3.7 Activities

To answer questions online and to receive **immediate feedback** and **sample responses** for every question, go to your learnON title at www.jacplus.com.au. *Note:* Question numbers may vary slightly.

Remember

1. How do groynes and sea walls help to manage or protect a coastal landscape?
2. Draw a diagram to explain your answer to question 1. Refer to figure 3 in subtopic 3.4 to help with your diagram.

Explain

3. Describe what will happen to Adelaide's southern beaches if they stop being replenished with trucks of sand.

Discover

4. Research another example of coastal landscape management. Identify why the management strategies were put in place and comment on their success. Examples of *places* that would be good to research include Cape Woolamai, the Gold Coast, Melbourne bayside beaches, Polder coastline of the Netherlands, Bondi, Cottesloe, Venice Beach or Waikiki.

Think

5. Imagine that you own a holiday house that is built on coastal dunes within 15 metres of the beach. After a powerful storm, the beach in front of your house is eroded and your house is now only five metres from the sea. What are your options? Work out a series of strategies that you could implement which may save your house from falling into the sea. Include diagrams to illustrate your plan.
6. Identify the strengths and weaknesses, for your house and your neighbours' houses, of the management proposal you created to answer question 5.
7. (a) Refer to figure 4. Describe the *changes* that have occurred to Adelaide's coastline over the past 7000 years.
   (b) Refer to figures 3 and 5. Describe the *changes* the Living Beaches Strategy has made to the Adelaide coastline and the reasons for these *changes*.

 Deepen your understanding of this topic with related case studies and questions.
❯ **Managing coasts**

3.8 How do I undertake coastal fieldwork?

3.8.1 Your fieldwork task

The best way to understand the physical processes and human activities that affect a specific coastal landscape is to visit it. A fieldwork activity will allow you to put the knowledge you have gained in the classroom into practice. Your fieldwork will also allow you to enjoy the coastal landscape in magnificent 3D.

Any coastal landscape would be suitable to investigate. Once a fieldwork site has been identified, there is quite a lot of planning that you should do before you get there.

What is your fieldwork task?

Your task is to identify the landforms and dynamic nature of a coastal landscape and to recognise and assess the influence of people on it.

In class

1. Prepare a base map of the fieldwork site or sites. On this base map, mark in the location of the coastal landscape's natural features (such as beach, rock, dunes, water, vegetation) and **human features** (such as seawall, groyne, steps, lawn, shelter, jetty). Using Google Maps or a topographic map is an excellent way of identifying the specific details of the coastal landscape.
2. Looking at the aerial shot on Google Maps will also allow you to see the pattern of the waves as they move to the shore. Does it look as if longshore drift is occurring on the day this image was taken?

On your field trip

What do you need to do at the coast to collect your information?

It is good to work in groups to collect your data in the field. It is then possible for some students to take measurements and some to record. Sharing tasks means that there will be others with whom to discuss what you have recorded. On returning to class you can pool your observations. You will need recording sheets, pencils, a digital or phone camera, tape measure, compass and maybe a clinometer. You could also collect information using data logging equipment, a GPS locator, weather recording equipment and notepads. Your group should decide what equipment is the most practical and relevant for collecting the data you need.

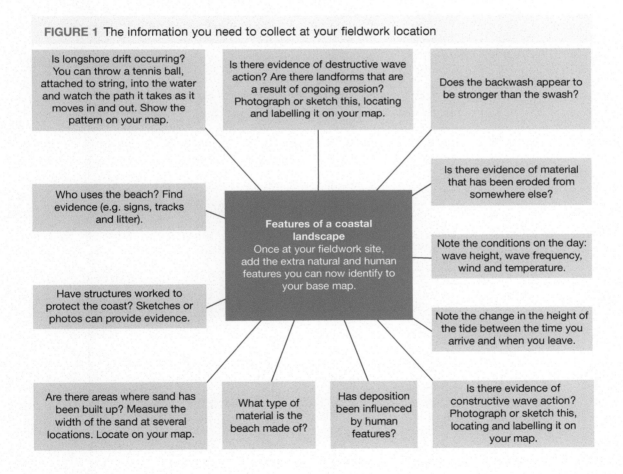

FIGURE 1 The information you need to collect at your fieldwork location

Is longshore drift occurring? You can throw a tennis ball, attached to string, into the water and watch the path it takes as it moves in and out. Show the pattern on your map.

Is there evidence of destructive wave action? Are there landforms that are a result of ongoing erosion? Photograph or sketch this, locating and labelling it on your map.

Does the backwash appear to be stronger than the swash?

Who uses the beach? Find evidence (e.g. signs, tracks and litter).

Is there evidence of material that has been eroded from somewhere else?

Features of a coastal landscape
Once at your fieldwork site, add the extra natural and human features you can now identify to your base map.

Note the conditions on the day: wave height, wave frequency, wind and temperature.

Have structures worked to protect the coast? Sketches or photos can provide evidence.

Note the change in the height of the tide between the time you arrive and when you leave.

Are there areas where sand has been built up? Measure the width of the sand at several locations. Locate on your map.

What type of material is the beach made of?

Has deposition been influenced by human features?

Is there evidence of constructive wave action? Photograph or sketch this, locating and labelling it on your map.

You may not be able to return to your fieldwork site, which means your data needs to be very detailed.

- Always record the location of the information on your map.
- Take photos of the coastal landscape, including the landforms and human structures.
- Measure distances and heights.
- Draw **field sketches** to remind you of details. Even when you have photographed something, a field sketch allows you to annotate the diagram so that you can remember important characteristics about how it was formed or

FIGURE 2 Investigating the rocky shores of a coastal landscape

the direction of longshore drift. Do not worry if you are not a gifted artist, as there are apps that allow you to convert your photos to sketches when you get back to class.

FIGURE 3 Field sketch

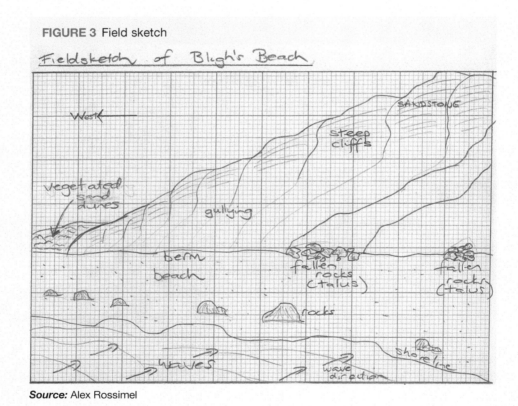

Source: Alex Rossimel

Back in class

Now that you have collected your information in the field, you need to present your findings about the coastal landscape you visited.

There are many ways that you could present this information. Your fieldwork report could be presented as a poster, website, PowerPoint presentation, booklet, blog, movie, news report or podcast. Consider using Google Maps and uploading images of the sites you visited. You will need to present the data you collected and describe your findings.

FIGURE 4 Students on a fieldwork trip, measuring the slope of a sandy beach.

3.9 SkillBuilder: Constructing a field sketch

WHAT ARE FIELD SKETCHES?

Field sketches are drawings completed during fieldwork — geography outside the classroom. Field sketches allow a geographer to capture the main aspects of landscapes in order to edit the view, focusing on the important features and omitting the unnecessary information.

Go online to access:

- a clear step-by-step explanation to help you master the skill
- a model of what you are aiming for
- a checklist of key aspects of the skill
- a series of questions to help you apply the skill and to check your understanding.

FIGURE 1 Field sketch of Cape Schanck

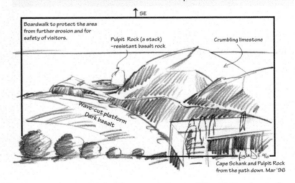

Source: © Geography Teachers' Association of Victoria Inc. *Interaction*, journal of the GTAV, June 1998. Illustration redrawn by Harry Slaghekke.

3.10 What's in a rip?

3.10.1 What is a rip current?

The movement of water within the coastal landscape can be hazardous to people as well as to coastal landforms.

If you have ever walked along a surf beach, you know about the uneven power of waves as the water returns to the ocean. This backwash can either be a gentle, non-threatening flow or it may be so powerful that it can be difficult for you to remain standing. The powerful flow of water returning to the sea is called **undertow**. If the seabed adjacent to the beach has an even slope, then the undertow is relatively harmless. Water returns to the sea and washes ashore in the next wave. Undertow can also take material from the beach and deposit it in the wave zone. If the seabed adjacent to the beach has an uneven topography, or shape, the undertow will be more threatening.

Where the flow of an undertow is concentrated, as occurs in a gap between offshore sandbars, or in a deeper seabed channel, this may result in the formation of a **rip current**. A rip current, or rip, is a strong current that runs offshore from the beach. If you are caught in a rip, it feels as if you are in a fast-moving river channel being whisked out to sea. The fast-moving water erodes a deep channel in the sea bed.

3.10.2 How do you identify a rip?

Before you swim at a surf beach, it is important to try to identify the location of possible rips. If the beach is patrolled by surf lifesavers, the yellow and red flags will be placed in a rip-free part of the beach. Looking at figures 1 and 3, you can see that the sea where the rip current is found is actually different from the sea on either side of it. Typically, where there is a rip current, the sea looks calmer, with surface ripples indicating a flow of water away from the beach. The waves are not as big and the water may be a

FIGURE 1 Rips at Bondi Beach, Sydney. Rips are often easier to identify from a height.

Rips

FIGURE 2 A rip current highlighted with purple dye

Longshore currents

Breaker zone

Return flow

Rip head

Shallow water

Deep water

Rip current

FIGURE 3 A rip current can take you offshore very quickly, although it will not carry you far out to sea.

Longshore drift

Shallow water

Rip current (deep water)

Longshore drift

Shallow water

Rip head

darker colour because it is deeper. The calmer water often tricks swimmers into thinking that this is a safer place to swim.

3.10.3 What should you do if you are caught in a rip?

Rips do not actually pull you under the water, but being caught in a rip current causes many people to drown on our beaches each year. Figure 3 shows that a rip will not take you far out to sea, although it will take you offshore very quickly. If the waves are large, then the speed of the rip will be faster and more dangerous.

If you are caught in a rip, the most important thing to do is not to panic. Do not try to swim against the rip, because even the strongest swimmer will get tired. Most drownings at Australian beaches occur when the beach is not being patrolled. This means that the best way to avoid the hazard created by a rip current is to swim between the flags at a patrolled surf beach.

FIGURE 4 Rip warning poster

3.10 Activities

To answer questions online and to receive **immediate feedback** and **sample responses** for every question, go to your learnON title at www.jacplus.com.au. *Note:* Question numbers may vary slightly.

Remember

1. How is undertow different from a rip current?
2. What colour are the flags used by lifesavers to identify the safe part of a beach for swimming?

Explain

3. Explain why a beach that has large waves would have more powerful rip currents.
4. Using the information illustrated in figure 4, make a list of five things that you should do to ensure your safety if you are caught in a hazardous rip current.
5. If you are caught in a rip, explain why you should not try to swim towards the shore. Annotate a diagram of a rip to explain your answer.

Discover

6. True or false? Rip currents tend to have less wave activity than surrounding areas.
7. What are the features of a rip current that make them potentially dangerous to swimmers?
8. Download several images of surf beaches from the net and annotate them to identify potential rip currents. Give reasons for the *places* you marked as rips.

Think

9. Design a rips awareness campaign to warn swimmers about the potential hazards caused by rip currents. It may be a campaign for television, radio or newspapers. Select the most effective campaign strategies and, as a class, make it your task to promote rip current awareness within your school.
10. Describe the *scale* of the rips in figure 1 in relation to the whole Bondi Beach region.

---

learn on RESOURCES — ONLINE ONLY

✦ **Try out this interactivity:** Ripped out (int-3105)

3.11 How does water form river landscapes?

3.11.1 Moving water

Erosion, transportation and deposition by running water are the main processes that create our landscapes. Some rivers, such as the Gordon River in Tasmania, are **perennial**; some, such as Coopers Creek in Queensland, are **intermittent**; others, such as the Colorado River in the United States, have eroded amazing landforms like the Grand Canyon.

Water is always on the move. It evaporates and becomes part of the water cycle; it rains and flows over the surface of the Earth and into streams that make their way to a sea, lake or ocean; and it soaks through the pores of rocks and soil into **groundwater**.

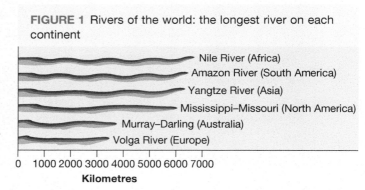

FIGURE 1 Rivers of the world: the longest river on each continent

- Nile River (Africa)
- Amazon River (South America)
- Yangtze River (Asia)
- Mississippi–Missouri (North America)
- Murray–Darling (Australia)
- Volga River (Europe)

0 1000 2000 3000 4000 5000 6000 7000
Kilometres

3.11.2 River systems and features

A river is a natural feature, and what we see is the result of the interaction of a range of inputs and processes. All parts of the Earth are related to the formation of river landscapes. This includes the lithosphere (rocks and soil), the hydrosphere (water), the biosphere (plants and animals) and the atmosphere (temperature and water cycle). Changes can happen quickly and over a very long period of time. Changes at one location along a river can have an effect at other locations along the river.

Water flows downhill, and the source (the start) of a river will be at a higher altitude than its mouth (the end). As the water moves over the Earth's surface, it erodes, transports and deposits material.

The volume of water and the speed of flow will influence the amount and type of work carried out by a river. A fast-flowing flooded river will erode enormous amounts of material and transport it **downstream**. As the speed or volume of the water decreases, much of the material it carries will be deposited.

FIGURE 2 A river system

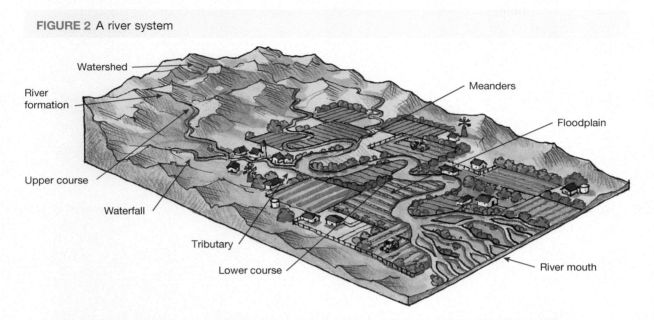

FIGURE 3 (a) Long profile of a river — a view along its length. The slope of the river tends to get flatter and the riverbed gets smoother as it moves downstream. (b)–(d) Cross-sections showing the shape of the river channel and valley at three points along the river. Arrows indicate the main direction of erosion.

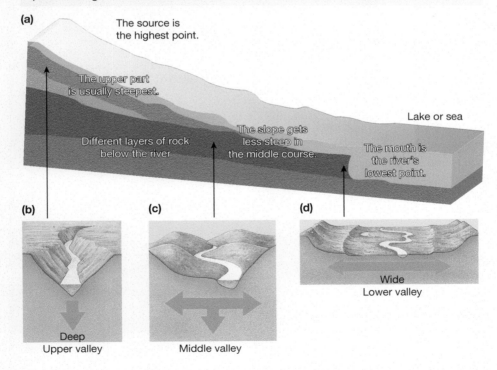

Watershed

A river gathers its water from a region known as its drainage basin, or **catchment** (figure 4). The boundary of this region is identified by mountain tops, hilltops or any land that is slightly higher than surrounding land. This is known as the watershed, and it is the point that divides the direction of water flow.

River formation

Even the biggest rivers begin with water from rain or melting snow in mountains or hills. This water has collected in tiny depressions called rills. These rills grow larger when they collect more water, and when they combine they begin to look like streams. Many streams contribute to a river (see figure 2).

FIGURE 4 The watershed and catchment, or drainage basin of a river system

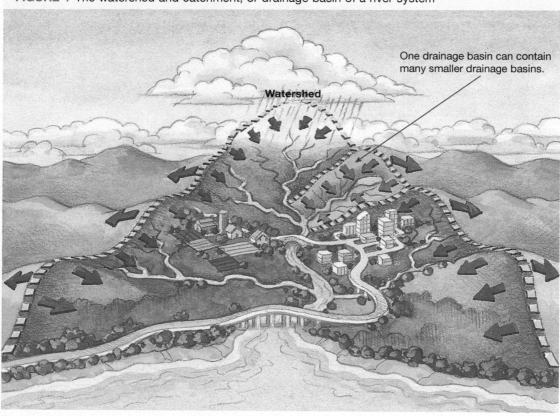

One drainage basin can contain many smaller drainage basins.

Watershed

Source: Adapted from an image by RecycleWorks www.RecycleWorks.org

Upper course

Waters in a river channel flow over steeper slopes in the upper reaches. The force of rushing water on a steep slope cuts downwards and creates a V-shaped valley. The river then tends to follow a fairly straight course (see figure 3).

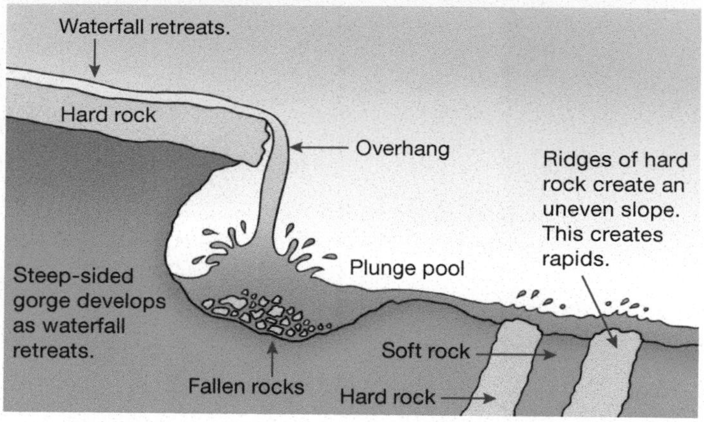

Waterfall

When a river meets resistant rock, a waterfall can occur (see figure 5). If the river has to cross bands of resistant rock, rapids will form. The turbulent water flow in rapids is called white water. A plunge pool forms at the base of a waterfall when rocks and soil moved by the fast flowing water erode the banks and base of the river.

FIGURE 5 A waterfall

Waterfall retreats.

Hard rock

Overhang

Ridges of hard rock create an uneven slope. This creates rapids.

Steep-sided gorge develops as waterfall retreats.

Plunge pool

Fallen rocks

Soft rock

Hard rock

Tributary

A river or stream that adds or contributes water to the main river is known as a **tributary** (see figure 6). The place where two rivers join is called the confluence.

Meanders

On flatter land, a river is wider than it is in the hills, and water added from tributaries has increased its volume (see figure 3). Much of the erosion is in a sideways direction, and the valley of a river is much wider. Sideways erosion causes meanders (curves) along its course (see figure 7). Over time, a meandering river will change the path it follows, as some bends become more obvious and some disappear. A meander that is cut off is called an oxbow lake.

Floodplains

Flooding over thousands of years creates floodplains. During a flood, the water flows over the banks of the river. Once outside the river, it slows down and deposits the alluvium it was transporting. This alluvium is often very fertile (see figure 2). These regions are highly suitable for farming and settlement (see figure 8).

River mouth

Deltas are found at the mouths of large rivers, like the Mississippi. A delta is formed when the river deposits its material faster than the sea can remove it. The material is a mix of mud, sand and clay. The river will sometimes split up into smaller streams to find its way through the deposited material. These little streams are called distributaries. We recognise three main shapes of delta: fan shaped, arrow shaped and bird-foot shaped. The shape is influenced by tides, ocean waves and the volume of sediment and river water.

Sometimes a river will have a wide mouth, where fresh water and salt water mix. This is known as an estuary.

FIGURE 6 Tributaries of the Darling River which, in turn, is a tributary of the Murray River

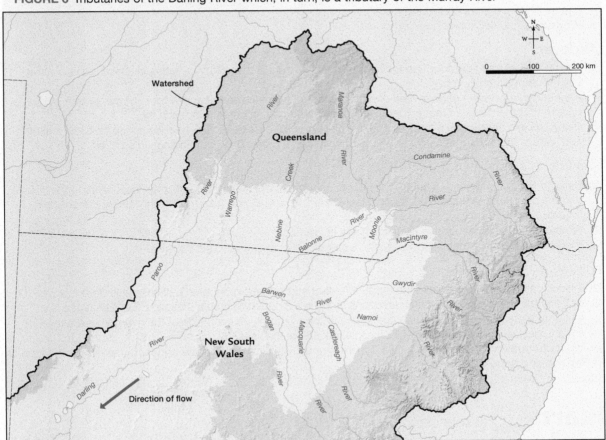

Source: Spatial Vision

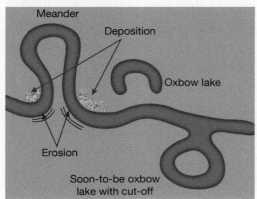

FIGURE 7 The formation of a meander and oxbow lake

Meander
Deposition
Oxbow lake
Erosion
Soon-to-be oxbow lake with cut-off

FIGURE 8 Murray River meanders

3.11 Activities

To answer questions online and to receive **immediate feedback** and **sample responses** for every question, go to your learnON title at www.jacplus.com.au. *Note:* Question numbers may vary slightly.

Remember

1. Refer to figure 1 and compare the *scale* of Australia's longest river with the world's longest river.
2. Sketch the long profile shown in figure 3 and label the source, the mouth and the direction the river flows.
3. What feature, other than water, has to be present for waterfalls and rapids to form? Refer to figure 5.

Explain

4. Explain how rivers are part of the water cycle.
5. Why do people settle and farm on floodplains?

Discover

6. Identify a river that flows through the capital city in one state or territory in Australia. Describe its source, any tributaries, and its mouth.
7. After some rain, investigate an area of bare ground on a small slope near school or home. Sketch the pattern that the rills have made. Identify the watershed and catchment for each rill.
8. Using Google Earth or an atlas, find the Nile delta, the Ebro delta and the Mississippi delta. Draw a sketch and write a short description of the shape of each delta, presenting your findings in a table.

Predict

9. Refer to figure 8. Sketch a diagram to show the course of the meandering Murray River. Mark in the course that the river used to take. Predict and label where the next oxbow lake might form. Show the possible future course of the river.
10. What do you think will happen to deltas if sea levels rise?
11. Predict the *changes* that will occur to the waterfall in figure 5. Justify your answer.

Think

12. Produce a flowchart or animation to explain the formation of an oxbow lake, a delta, a waterfall or rapids.
13. Refer to the paragraph about meanders and to figures 3 and 7. Sketch a cross-section (like figure 3) of a river at a meander. This will show the shape of the riverbank on each side of the river. What are the advantages and disadvantages of living on each side of the river?
14. What *changes* will occur along a river if there is unusually high rainfall in its upper course? Think in terms of erosion and deposition.

learnon RESOURCES — ONLINE ONLY

Try out this interactivity: River carvings (int-3104)

3.12 How are river landscapes managed?

3.12.1 Mississippi River

Rivers are vital. Plants and animals depend on their waters for survival. People also rely on rivers for their waters and have diverted rivers for flood control, irrigation, power generation, town water supplies, waste disposal and recreation.

The mighty Mississippi River is approximately 3700 kilometres long and is the second longest in the United States. It flows through 10 states (see figure 1). The drainage basin, or catchment, for the river covers 40 per cent of the country, and includes all or part of 31 states and two Canadian provinces. The drainage system is made up of thousands of rivers and streams, including the Missouri.

Importance of the river

The Mississippi has been a major contributor to the economic growth of the United States.

- It is important for transporting goods, such as fuel, coal, gravel, chemicals, steel, cement and farm produce. The **barges** on the river are able to connect to ocean shipping at Baton Rouge in Louisiana.
- It supplies water for cities and industries and irrigation for farming.
- Much of its floodplain has been cleared for farmland.
- The river basin also supports natural biodiversity. It has many species of mussels, 25 per cent of all fish species in North America, and over 300 species of birds that use the river during migration and breeding.

FIGURE 1 The Mississippi drainage basin

Source: Spatial Vision

Floods

The river has created the geographical characteristics that have always attracted settlement. The source of the river is at an altitude of 450 metres above sea level, and the river drops in altitude very quickly. The last 1000 kilometres of the river's journey is through a wide floodplain that is the result of many floods over hundreds of years. Under natural conditions, the river had high water levels in early spring and much lower levels by early autumn.

Floods are a major issue for businesses, homes and farms. There have been many significant floods; for instance, in 1849, 1850, 1882, 1912, 1913, 1927, 1983, 1993 and 2011. After the floods of 1927, the Mississippi River and Tributaries Project was set up with the goal of preventing destructive floods and keeping the river open for navigation.

River management

The Mississippi River and Tributaries Project uses many strategies to manage the river. The aim is to satisfy the needs of farming, towns, industry, transport and **ecosystems**. There are many dams to control water levels in the river.

Management issues

- The strategies are expensive.
- Continuous dredging is needed.
- Levees are being built higher — some now seven metres high — and it is hard work to make sure they don't leak or break.
- Water is powerful and the river still wears away at weak points along the banks.
- If a levee breaks or if water goes over the top, flood damage can be very bad.
- The floodplain does not receive much sediment from the river.
- The river water is not as clean as it used to be.
- Natural habitats are damaged by dredging or concreting.
- The delta is decreasing in size.

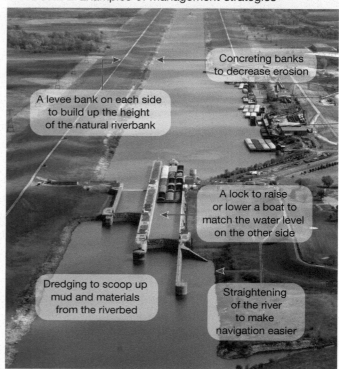

FIGURE 2 Examples of management strategies

Concreting banks to decrease erosion

A levee bank on each side to build up the height of the natural riverbank

A lock to raise or lower a boat to match the water level on the other side

Dredging to scoop up mud and materials from the riverbed

Straightening of the river to make navigation easier

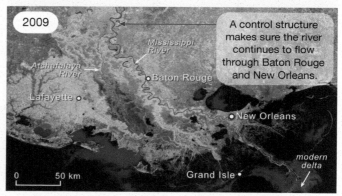

FIGURE 3 Predicted changes to the Mississippi Delta, 2009 to 2100

2009

Mississippi River

Atchafalaya River

Lafayette

Baton Rouge

New Orleans

Grand Isle

modern delta

A control structure makes sure the river continues to flow through Baton Rouge and New Orleans.

0   50 km

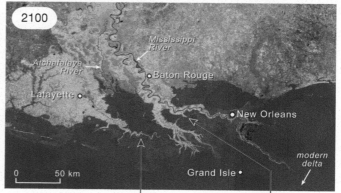

2100

Mississippi River

Atchafalaya River

Lafayette

Baton Rouge

New Orleans

Grand Isle

modern delta

0   50 km

The river water contains too many nutrients from towns and farms. This is a concern for the fishing industry.

Less sediment is reaching the delta; the delta is becoming smaller.

3.12 Activities

To answer questions online and to receive **immediate feedback** and **sample responses** for every question, go to your learnON title at www.jacplus.com.au. *Note:* Question numbers may vary slightly.

Remember

1. Refer to figure 1 and name key tributaries of the Mississippi River. In which general direction does the Mississippi flow from its source to its mouth?

Explain

2. Why is the river important to the United States? Classify each reason as one or a combination of the following: social, economic or *environmental*.

Discover

3. How close will Baton Rouge be to the sea in 2100?
4. Use the **Land loss in the Mississippi delta** weblink in the Resources tab. Which parts of the delta have been most affected by erosion?
5. Use the **Mississippi River watershed** weblink in the Resources tab to watch a video about the river. What do you notice about the *scale* of the watershed and the location of the Mississippi River?

Predict

6. What do you think would be the main management strategies on the Mississippi River during a year of heavy rainfall? What do you think would be the main management strategies during a drought?

Think

7. Do you agree or disagree with the following statement? 'A strategy implemented in one part of the river will have an impact on another part of the river.' As you find evidence from this topic, place it in a table, or under subheadings. Write a conclusion based on your findings.

 RESOURCES — ONLINE ONLY

Explore more with these weblinks: Land loss in the Mississippi delta, Mississippi River watershed

3.13 Which landscapes are formed by ice?

Access this subtopic at **www.jacplus.com.au**

3.14 Do all rivers flow to the sea?

3.14.1 Okavango River

Some rivers do not reach the sea. The Okavango River in Africa has created an inland delta, its water finally disappearing into ancient salt pans.

This perennial river has its source in the highlands of Angola, flows roughly south-east for about 1000 kilometres, passing along the Namibian border, to its delta in Botswana. Very little rain is added to the river in Botswana and it takes about a month for the water to flow from its source to the delta. The delta doubles in size when it floods in the dry season. This makes it an oasis in the Kalahari Desert.

The delta region contains many islands and is fairly flat. When the volume of the river increases, the water slowly spreads out with some of the water making it as far as the Makgadikgadi Salt Pans. A lot of water is lost from the delta through very high rates of evaporation.

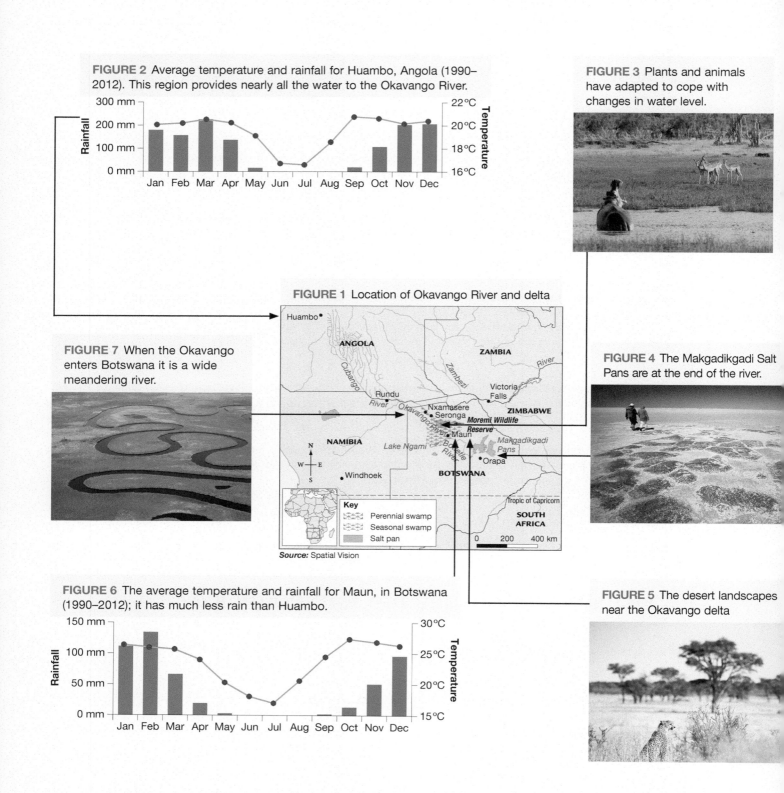

FIGURE 2 Average temperature and rainfall for Huambo, Angola (1990–2012). This region provides nearly all the water to the Okavango River.

FIGURE 3 Plants and animals have adapted to cope with changes in water level.

FIGURE 1 Location of Okavango River and delta

Key
Perennial swamp
Seasonal swamp
Salt pan

Source: Spatial Vision

FIGURE 7 When the Okavango enters Botswana it is a wide meandering river.

FIGURE 4 The Makgadikgadi Salt Pans are at the end of the river.

FIGURE 6 The average temperature and rainfall for Maun, in Botswana (1990–2012); it has much less rain than Huambo.

FIGURE 5 The desert landscapes near the Okavango delta

3.14.2 Diamantina River

CASE STUDY

The Diamantina River flows into Kati Thanda–Lake Eyre

The Diamantina River in Australia is an intermittent river about 900 kilometres long, with its source in Queensland and its mouth in inland South Australia at Kati Thanda–Lake Eyre. It is one of the rivers in the catchment of Kati Thanda–Lake Eyre. Some local rain adds to the water in the lake, but most of the water comes from cyclones and tropical rain in northern Queensland. Water can take three to 10 months to flow all the way to the lake. There are a few permanent waterholes along the river and there can be many dry years between the short periods when the river flows.

The temperatures in the catchment are high and only a small fraction of the Queensland rain reaches the lake. Water seeping into the soil also reduces the level of the lake. About once every eight years there is significant water in the lake. Kati Thanda–Lake Eyre is in an arid environment. The wet and dry periods influence the patterns of life of flora and fauna.

FIGURE 9 The river divides into many small channels.

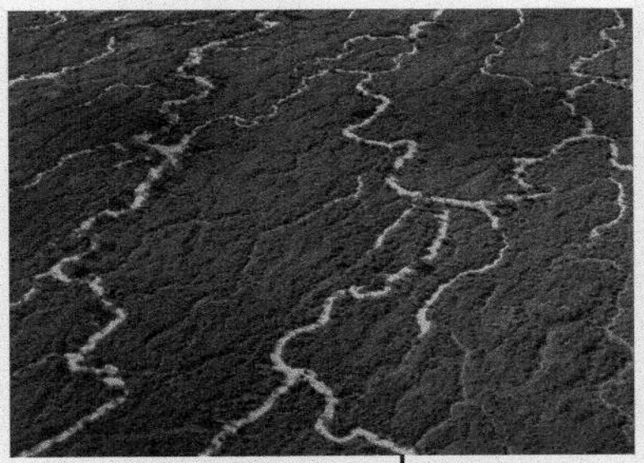

FIGURE 10 Average rainfall and temperature at Swords Range (1990–2012), the source of the Diamantina River

Jun: 19.33 mm

FIGURE 14 High temperatures lead to high evaporation of water, leaving a very salty mixture behind.

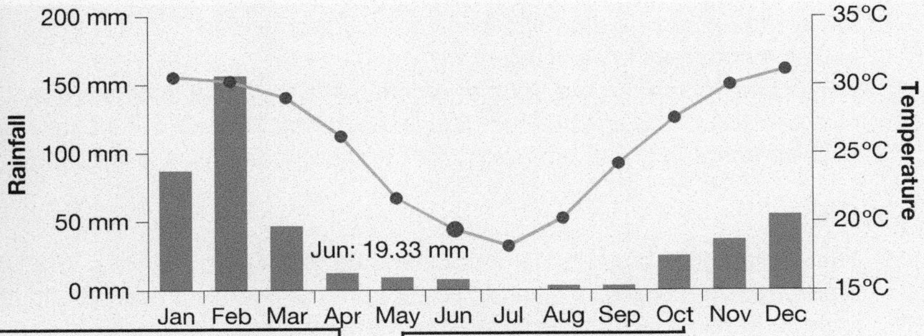

FIGURE 8 The location of the Diamantina River

Northern Territory

Tropic of Capricorn

Mount Isa

Queensland

Longreach

Charleville

Kati Thanda / Lake Eyre

South Australia

New South Wales

Key
Kati Thanda / Lake Eyre drainage basin
State border

0   200   400 km

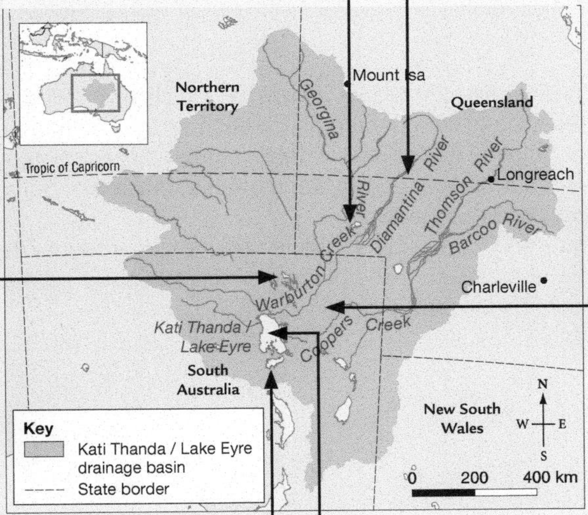

Source: Geoscience Australia

FIGURE 11 The desert environment surrounding the lake

FIGURE 13 Birds fly long distances to breed and feed on the fish life in the water.

FIGURE 12 Average temperature and rainfall at Kati Thanda–Lake Eyre (1990–2012). The lake is in a hot, dry region.

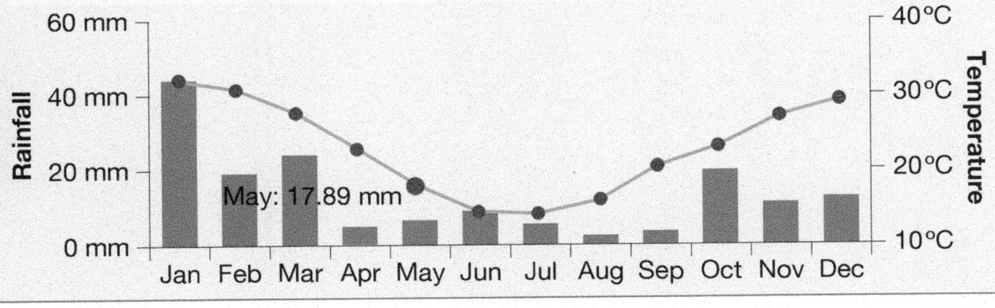

May: 17.89 mm

3.14 Activities

To answer questions online and to receive **immediate feedback** and **sample responses** for every question, go to your learnON title at www.jacplus.com.au. *Note:* Question numbers may vary slightly.

Remember

1. What is the difference between the average annual rainfall at the source and the mouth of the Diamantina River?
2. Describe the distribution of rain in a year at Huambo and at Swords Range.
3. Refer to figures 7 and 9 and describe the difference in the shape of the river channel.

Explain

4. Distinguish between a perennial river and an intermittent river.
5. How is it possible that the Okavango delta floods in the dry season?
6. Refer to figures 3 and 13 and compare the flora and fauna at the mouth of each river.
7. How does high temperature lead to water loss?

Discover

8. In which hemisphere are the Okavango and Diamantina Rivers?
9. In which continent is each river?
10. What is a cyclone? Name another *place* in Australia that is affected by cyclones.
11. Find out about one species that flourishes at Kati Thanda–Lake Eyre when the lake has water. How has this species adapted to the *environment*? What other species is it relying upon?

Predict

12. The Okavango is the only permanent source of water in Namibia. What would happen to the delta *environment* if Namibia diverted water from the river as it flows along its border? Think about the possible impact on people, on the *environment* and on the economy.
13. *Changes* to the climate are likely to make the Kati Thanda–Lake Eyre region hotter and drier. What impact will this have on the *environment* of the lake and the native species that flourish there?

Think

14. Compare the Okavango River and the Diamantina River. Present your findings in a visual manner to show similarities and differences. Consider characteristics such as the location of the river, length of river, rainfall at source, rainfall at mouth, perennial or intermittent nature, *environment* surrounding mouth, speed of flow, evaporation rate at mouth, water added along course, fauna at mouth, flora at the mouth, shape of the river channel and final end point of the river.
15. The Okavango Delta and Kati Thanda–Lake Eyre are both popular tourist destinations. Provide two reasons why you think people are attracted to these places. Which one has a permanent tourism industry? How is this possible? When would be the best time to visit each location?

3.15 SkillBuilder: Reading contour lines on a map

WHAT ARE CONTOUR LINES?

Contour lines drawn on the map join all places of the same elevation (height) above sea level. Contour maps are used to show the relief (shape) of the land and the heights of the landscape. Maps with contour lines show the relief of the land and help people to identify features.

Go online to access:

- a clear step-by-step explanation to help you master the skill
- a model of what you are aiming for
- a checklist of key aspects of the skill
- a series of questions to help you apply the skill and to check your understanding.

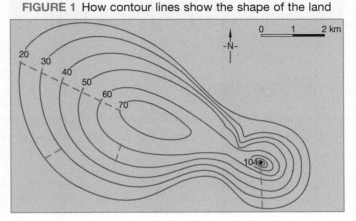

FIGURE 1 How contour lines show the shape of the land

3.16 Review

3.16.1 Review

The Review section contains a range of different questions and activities to help you revise and recall what you have learned, especially prior to a topic test.

3.16.2 Reflect

The Reflect section provides you with an opportunity to apply and extend your learning.

Access this subtopic at **www.jacplus.com.au**

TOPIC 4
Desert landscapes

4.1 Overview

Numerous **videos** and **interactivities** are embedded just where you need them, at the point of learning, in your learnON title at www.jacplus.com.au. They will help you to learn the content and concepts covered in this topic.

4.1.1 Introduction

Approximately one-third of the Earth's land surface is desert — arid land with little rainfall. These arid regions may be hot or cold. The actions of wind and, sometimes, water shape the rich variety of landscapes found there.

Arches in a desert landscape, Utah, United States

Starter questions

1. Have you ever visited a desert? Where was it? Why did you go?
2. Make a list of the geographical characteristics of desert areas. What is the most common feature of a desert?
3. Find a description of a desert online. Copy the text and use the **Wordle** weblink in the Resources tab to create a Wordle. What is the most common word in your Wordle? Why do you think this is the case?
4. Use the **Sahara** weblink in the Resources tab to watch a video of desert sandstorms.
   (a) Describe the sandstorm you see. Would it be difficult for people to cope with such an event?
   (b) Describe how some animals are adapted to cope with desert sandstorms.
   (c) How do the wind and sand shape the desert?

INQUIRY SEQUENCE

learn on RESOURCES — ONLINE ONLY

🔗 **Explore more with these weblinks:** Wordle, Sahara

4.2 What is a desert?

4.2.1 Defining a desert

A desert is a hot or cold region with little or no rainfall. Around one-third of the Earth's surface is desert and is home to about 300 million people.

Although they receive little rainfall, most deserts receive some form of precipitation. When it does rain, it is usually during a few heavy storms that last a short time.

TABLE 1 Types of deserts

Rainfall (mm/year)	Type of desert	Examples
< 25	Hyper-arid	Namib; Arabian
25–200	Arid	Mojave
200–500	Semi-arid	Parts of Sonoran Desert

4.2.2 Hot deserts

Most of the world's hot deserts are located between the Tropic of Cancer and the Tropic of Capricorn (see figure 3). They have very hot summers and warm winters. Temperature extremes are common, because cloud cover is rare and humidity is very low; this means there is nothing to block the heat of the sun during the day, or prevent its loss at night. Temperatures can range between around 45 °C and –15 °C in a 24-hour period.

FIGURE 1 The Sahara, an example of a hot desert

4.2.3 Cold deserts

Cold deserts lie on high ground generally north of the Tropic of Cancer and south of the Tropic of Capricorn (see figure 3). They include the polar deserts. Any precipitation falls as snow. Winters are very cold and often windy; summers are dry and cool to mild.

FIGURE 2 The Gobi, an example of a cold desert

4.2.4 Deserts of the world

FIGURE 3 The distribution of the world's deserts

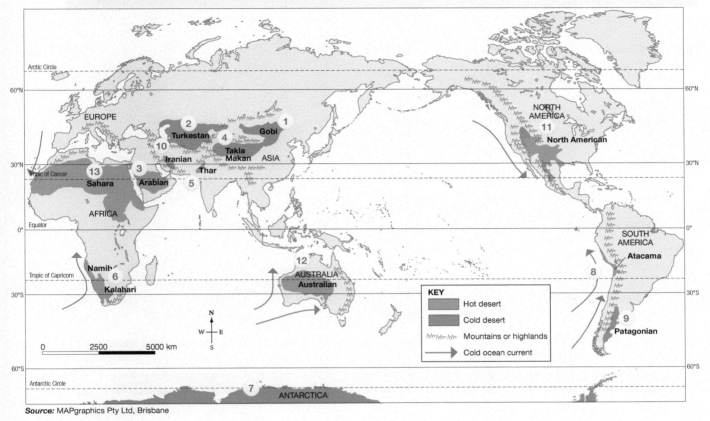

Source: MAPgraphics Pty Ltd, Brisbane

1 **Gobi Desert:** Asia's biggest desert, the Gobi, is a cold desert. It sits some 900 metres above sea level and covers an area of some 1.2 million square kilometres. Its winters can be freezing.

2 **Turkestan Desert:** The cold Turkestan Desert covers parts of south-western Russia and the Middle East.

3 **Arabian Desert:** This hot desert is as big as the deserts of Australia. Towards its south is a place called Rub al-Khali (meaning 'empty quarter'), which has the largest area of unbroken sand dunes, or erg, in the world.

4 **Takla Makan Desert:** The Takla Makan Desert is a cold desert in western China. Its name means 'place of no return'. The explorer Marco Polo crossed it some 800 years ago.

5 **Thar Desert:** The Thar Desert is a hot desert covering north-western parts of India and Pakistan. Small villages of around 20 houses dot the landscape.

6 **Kalahari and Namib deserts:** The Namib Desert extends for 1200 kilometres down the coast of Angola, Namibia and South Africa. It seldom rains there, but an early-morning fog often streams across the desert from the ocean. The dew it leaves behind provides moisture for plants and animals. It joins the Kalahari Desert, which is about 1200 metres above sea level.

7 **Antarctic Desert:** The world's biggest and driest desert, the continent of Antarctica, is another cold desert. Only snow falls there, equal to about 50 millimetres of rain per year.

8 **Atacama Desert:** The Atacama Desert is the driest hot desert in the world. Its annual average rainfall is a tiny 0.1 millimetre.

9 **Patagonian Desert:** The summer temperature of this cold desert rarely rises above 12 °C. In winter, it is likely to be well below zero, with freezing winds and snowfalls.

10 **Iranian Desert:** Two large deserts extend over much of central Iran. The Dasht-i-Lut is covered with sand and rock, and the Dasht-i-Kavir, mainly in salt. Both have virtually no human populations.

11 **North American deserts:** The desert region in North America is made up of the Mojave, Sonoran and Chihuahuan deserts (all hot deserts) and the Great Basin (a cold desert). The Great Basin's deepest depression, Death Valley, is the lowest point in North America.

12 **Australian deserts:** After Antarctica, Australia is the driest continent in the world. Its deserts are generally flat lands, often vibrant in colour.

13 **Sahara Desert:** The largest hot desert in the world, the Sahara stretches some nine million square kilometres across northern Africa over 12 countries. Only a small part is sandy. It is the sunniest place in the world.

4.2 Activities

To answer questions online and to receive **immediate feedback** and **sample responses** for every question, go to your learnON title at www.jacplus.com.au. *Note:* Question numbers may vary slightly.

Remember

1. What climate conditions are needed for hot and cold deserts to form?
2. Where is the sunniest *place* in the world?
3. Name three deserts in the Asia–Pacific region.

Explain

4. Describe key differences between hot and cold deserts.

Discover

5. Look carefully at the map in figure 3, and read the text.
   (a) Which continent has the largest area of hot desert?
   (b) Which continent has the largest area of cold desert?
   (c) What is the largest hot desert in the world?
   (d) What is the largest hot desert in the Asia–Pacific region?
   (e) Which is the driest continent in the world?
   (f) Which continent contains the driest hot desert?
   (g) Which North American desert contains the lowest land on the continent?

Think

6. Use the information in this subtopic to design a quiz of 10 questions entitled 'Deserts of the world'. Test your friends and family.
7. Draw up and complete a table like the one below to show your understanding of the locations and features of desert *environments*. Look for photos on the internet.

Name of desert	Mountain range	Continent	Ocean current	Photos

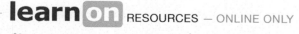

learn **on** RESOURCES — ONLINE ONLY

💧 **Try out this interactivity:** Great deserts of the world (int-3106)

4.3 SkillBuilder: Using latitude and longitude

WHAT IS LATITUDE AND LONGITUDE?

Latitude and longitude are imaginary grid lines encircling the Earth. The lines that run parallel to the equator are called **parallels of latitude** and are measured in degrees. Lines of longitude run from north to south from the North Pole to the South Pole. These are called **meridians of longitude** and are also measured in degrees. Lines of latitude and longitude are drawn on maps to help us locate places.

Go online to access:

- a clear step-by-step explanation to help you master the skill
- a model of what you are aiming for
- a checklist of key aspects of the skill
- a series of questions to help you apply the skill and to check your understanding.

FIGURE 1 Lines of latitude and longitude help us locate places on the Earth's surface.

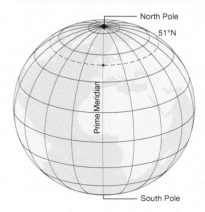

 RESOURCES — ONLINE ONLY

 Watch this eLesson: Using latitude and longitude (eles-1652)

 Try out this interactivity: Using latitude and longitude (int-3148)

4.4 What type of climate forms deserts?

4.4.1 The subtropics

Deserts form in many different parts of the globe: the subtropics; continental interior areas at middle latitudes; on the leeward side of mountain ranges; along coastal areas; and in the polar regions. The only common factor is their low rainfall — but why do these areas experience low rainfall?

Most of the world's greatest deserts are found in the subtropics near the Tropics of Cancer and Capricorn.

Because of the way the Earth rotates around the sun, areas around the equator receive more direct sunlight than anywhere else on Earth. This means the air there is always very hot. Hot air can hold much more moisture than cold air, so the **humidity** in these areas is always very high. (If you have ever visited

FIGURE 1 The formation of subtropical deserts in Africa

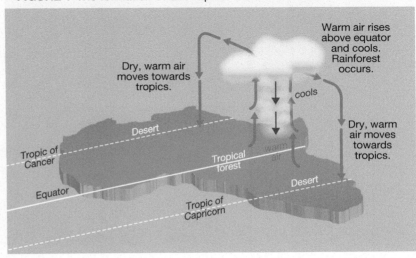

or live near a tropical rainforest or northern Australia, you will have experienced this hot humidity.) Hot air also rises. As the air heads upwards into the atmosphere above the equator, it drifts away, heading north and south.

The higher the air gets, the cooler it becomes. Cool air can't hold as much moisture, so it releases it as rain. Areas around the equator and to the immediate north and south of it (the tropics) receive frequent heavy downpours (see figure 1).

With its moisture gone, the cool, dry air continues moving north and south away from the equator until it meets zones of

FIGURE 2 The Sahara Desert of northern Africa is the world's largest and can experience temperatures as high as 57 °C.

high air pressure around the tropics. Here, it is forced downwards. The more the dry air descends, the warmer it gets. This means it can hold more moisture and it is likely to absorb any moisture that already exists in this environment. It is like using a sponge to wipe up some water on the kitchen bench; a dry sponge will absorb more of the spill than a wet sponge. This is how the subtropical deserts form.

Temperatures in these deserts are usually high all year round. In summer the heat is extreme, with daytime temperatures often going above 38 °C and sometimes as high as 49 °C. At night — with no clouds to provide insulation — temperatures drop quickly to an average of 21 °C in summer and sometimes below freezing during winter.

4.4.2 Rain-shadow deserts

Rain shadows form on the leeward side of a mountain range (opposite the windward side that faces rain-bearing winds). Deserts commonly form in rain shadows.

- Moist air blowing in from the ocean is forced to rise up when it hits a range of mountains. This cools it down. As cool air cannot hold as much moisture, it releases it as precipitation (see figure 3).

FIGURE 3 The formation of rain-shadow deserts

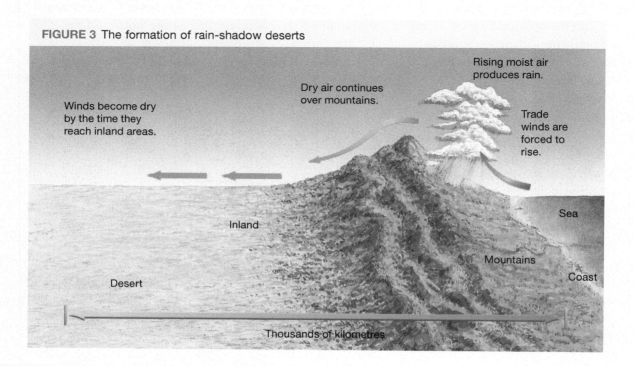

- By the time the air moves over the top of the range and down the other side, it is likely to have lost most, if not all, of its moisture. It will therefore be fairly dry.
- The more the air descends on the other side of the range, the more it warms up. Hence, it can hold more moisture. So, as well as not bringing any rain to the land, the air absorbs what little moisture the land contains.
- In time, as this pattern continues, the country in the rain shadow of the mountain range is likely to become arid.

An example of this is the Great Dividing Range in Australia; cool moist air produces winds on the eastern side of these mountains and desert to the west. The Mojave Desert in the south-western United States is located on the leeward side of the Sierra Nevada mountain range (figure 4).

4.4.3 Coastal deserts

Currents in the oceans are both warm and cold, and are always moving. Cold currents begin in polar and temperate waters (with moderate temperatures), and drift towards the equator. They flow in a clockwise pattern in the northern hemisphere, and in an anticlockwise pattern in the southern hemisphere. As they move, they cool the air above them (see figure 5).

If cold currents flow close to a coast, they can contribute to the creation of a desert. This occurs because cold ocean currents cause the air over the coast to become stable, which stops cloud formation. If the cool air the currents create blows in over warm land, the air warms up; it can then hold more moisture. It is therefore not likely to release any moisture it contains unless it is forced up by a mountain range. Large coastal deserts, including the Atacama Desert in Chile (figure 6) and the Namib Desert in Namibia (figure 7), are formed in this way.

4.4.4 Inland deserts

FIGURE 4 The Mojave Desert, United States

FIGURE 5 The formation of coastal deserts

Fog or rain occurs at sea, or on the coast.

Dry air blows inland.

cooler air

cold ocean current

FIGURE 6 The Atacama Desert is a coastal desert in northern Chile in South America and is the driest desert in the world. It is located on the leeward side of the Chilean Coast Range. In some areas, only around one millimetre of rain falls every 5–20 years.

Some deserts form because they are so far inland that they are beyond the range of any rainfall. By the time winds reach these dry centres, they have dumped any rain they were carrying or have become so warm they cannot release any moisture they still hold. The air that enters such areas is usually extremely dry and the skies are cloudless for most of the year. Summer daytime temperatures can rise as high as those of subtropical deserts. In winter, however, temperatures are much lower. Average daily temperatures below freezing are common during winter.

Examples of inland deserts are the central deserts of Australia (see figure 8), the Thar Desert in northwest India and the vast Gobi and Takla Makan deserts of Central Asia.

FIGURE 7 The coastal Namib Desert

FIGURE 8 The Simpson Desert in central Australia

4.4.5 Polar deserts

Polar deserts are areas with a precipitation rate of less than 250 millimetres per year and an average temperature lower than 10 °C during the warmest month of the year. Polar deserts cover almost five million square kilometres of our planet and consist mostly of rock or gravel plains. Snow dunes may be present in areas where precipitation occurs. Temperatures in polar deserts often alternate between freezing and thawing, a process that can create patterned textures on the ground as much as five metres across.

FIGURE 9 Although covered in frozen water, Antarctica receives little rain and is therefore classified as a desert.

4.4.6 Desert climate

Temperature

One geographical characteristic of many deserts is the high temperature, which quickly evaporates any water that might be around. The Earth's highest recorded temperature — 56.7 °C — occurred at Greenland Ranch in Death Valley, California, United States on 10 July 1913.

During the summer of 1923–24, the semi-arid town of Marble Bar in Western Australia (average rainfall 361 mm per year) experienced temperatures of more than 37.8 °C for 160 days in a row, from 31 October 1923 to 7 April 1924. This set an Australian temperature record, and possibly a world record. During that summer the temperature at Marble Bar peaked at 47.5 °C on 18 January 1924.

Rainfall

Although low rainfall is a characteristic of deserts, rain does fall and violent storms can sometimes occur. A record 44 millimetres of rain once fell within three hours in the Sahara. Large Saharan storms may deliver up to one millimetre of rain per minute. Normally dry stream channels, called arroyos or wadis, can quickly fill after heavy rains, and flash floods make these channels dangerous.

Monthly data for rainfall and temperature can be used to create climographs for other desert locations such as Khormaksar in Yemen and Alice Springs in Australia (see table 1).

FIGURE 10 Temperature and rainfall data can be displayed clearly in a climograph, such as this one for Yuma, Arizona.

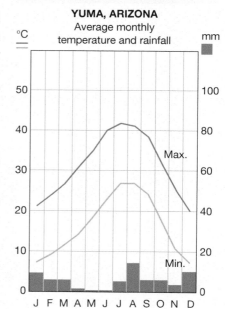

TABLE 1 Climate data for (a) Khormaksar, Yemen, and (b) Alice Springs, Australia

(a) Khormaksar, Yemen

	Jan	Feb	Mar	Apr	May	Jun	Jul	Aug	Sep	Oct	Nov	Dec	Total
Average temperature (°C)	25.0	25.5	27.0	28.5	30.5	33.0	32.0	32.0	32.0	28.5	26.5	25.5	
Average rainfall (mm)	5.0	0.0	5.0	0.0	0.0	0.0	5.0	3.0	0.0	0.0	0.0	5.0	23.0

(b) Alice Springs, Australia

	Jan	Feb	Mar	Apr	May	Jun	Jul	Aug	Sep	Oct	Nov	Dec	Total
Average temperature (°C)	28.5	27.7	24.8	20.0	15.4	12.4	11.5	14.3	18.3	22.8	25.8	27.7	
Average rainfall (mm)	40.5	41.5	34.7	16.6	17.0	16.7	12.1	10.0	9.0	20.0	25.3	37.2	280.6

4.4 Activities

To answer questions online and to receive **immediate feedback** and **sample responses** for every question, go to your learnON title at www.jacplus.com.au. *Note:* Question numbers may vary slightly.

Remember

1. Decide whether the following statements are true or false. Rewrite the false statements to make them true.
   (a) The cooler the air, the more moisture it can hold.
   (b) Rain shadows often contain dry areas of land.
   (c) Cold ocean currents cool the air above them.
   (d) Deserts do not form along coastlines.

Explain

2. Use figure 1 to explain why deserts form around areas near the tropics but not at the equator. Alternatively, form small groups and create a short drama performance to explain the process.
3. Use figure 3 and any other information in this subtopic to write a paragraph explaining why deserts tend to form in rain shadows. Alternatively, form small groups and create a short drama performance to explain the process.
4. Why do temperatures in deserts drop so much at night after being so high during the day?

Discover

5. Use the **Desert rain** weblink in the Resources tab to watch a video about desert rain, and then answer the following questions.
   (a) What is a flash flood?
   (b) What happens to water as it flows over sand? Think of what happens to water at the beach.
   (c) How do animals and plants respond to these rare water events?
   (d) Describe how the landscape quickly *changes* once there is water in the desert.

Think

6. Draw a diagram to explain how cold ocean currents influence the formation of a desert *environment* along the Western Australian coastline.
7. Use table 1 to draw climate graphs for Khormaksar, Yemen, and Alice Springs, Australia.

 **RESOURCES** — ONLINE ONLY

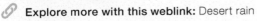 **Explore more with this weblink:** Desert rain

Try out this interactivity: How to make a desert (int-3107)

4.5 What are the processes that shape desert landforms?

4.5.1 Shaping the desert

Although most people imagine a sea of sand when they think of deserts, sand covers only about 20 per cent of the world's deserts. Sand is the end product of millions of years of erosion of other landforms such as rock and plateaus that, over time, are worn away by extremes of temperature, wind and water.

The landforms and patterns of a desert are created by a number of natural processes. The unprotected land surfaces are prone to erosion. After heavy rain, often a long distance from the desert flood plains, erosion of ancient river channels can be major. Extreme temperatures, along with strong winds and the rushing water that can follow a desert rainstorm, cause rocks to crack and break down into smaller fragments. This process is called weathering.

Erosional landforms

The process of erosion removes material such as weathered rock. Most erosion in deserts is caused by wind and, at times, running water. During heavy rainfall, water carves channels in the ground. Fast-flowing water can carry rocks and sand, which help to scour the sides of the channel. As vegetation is usually sparse or non-existent, there are few roots to hold the soil together. Eventually, deep gullies called wadis can form.

Erosion can also result from the action of wind and from chemical reactions. Some rock types, such as limestone, contain compounds that react with rainwater and then dissolve in it. Wind is a very important agent of transport and deposition, and can change the shape of land by abrasion — the wearing down of surfaces by the grinding and sandblasting action of windborne particles.

Depositional landforms

Materials carried along by rushing water and wind must eventually be put down. Over time these materials build up, forming different shapes and patterns in the desert. This process is called deposition.

Depositional landforms in deserts include alluvial fans, playas, saltpans and various types of sand dunes (see figure 1).

FIGURE 1 Desert landforms

1 A butte is the remaining solid core of what was once a mesa. It often is shaped like a castle or a tower.

2 Crescent-shaped barchan dunes are produced when sand cover is fairly light.

3 An arch, or window, is an opening in a rocky wall that has been carved out over millions of years by erosion.

4 An alluvial fan is the semicircular build-up of material that collects at the base of slopes and at the end of wadis after being deposited there by water and wind.

5 A playa lake may cover a wide area, but it is never deep. Most water in it evaporates, leaving a layer of salt on the surface. These salt-covered stretches are called saltpans.

6 Clay pans are low-lying sections of ground that may remain wet and muddy for some time.

7 The rippled surface on transverse dunes is the result of a gentle breeze blowing in the one direction.

8 An oasis is a fertile spot in a desert. It receives water from underground supplies.

9 A mesa is a plateau-like section of higher land with a flat top and steep sides. The flat surface was once the ground level, before weathering and erosion took their toll.

10 Sand dunes often start as small mounds of sand that collect around an object such as a rock. As they grow larger, they are moved and shaped by wind.

11 An inselberg is a solid rock formation that was once below ground level. As the softer land around it erodes, it becomes more and more prominent. Uluru is an inselberg.

12 A chimney rock is the pillar-like remains of a butte.

13 Star dunes are produced by wind gusts that swirl in from all directions.

14 Strong winds blowing in one direction form longitudinal dunes.

4.5.2 Sand dunes = depositional landforms

Different dune shapes are created by the action of the wind (see figure 2). These include crescent, linear, star, dome and parabolic. The most common are the crescent-shaped dunes that are formed when the wind blows in one direction (figure 3). They are usually wider than they are long and can move very quickly across desert landscapes.

Linear dunes are a series of dunes running parallel to each other. They can vary in length from a few metres to over 100 kilometres. It appears that winds blowing in opposite directions help create these dunes. The Simpson Desert in central Australia has linear dunes (figure 4).

FIGURE 2 The transport and deposition of sand creates and moves dunes.

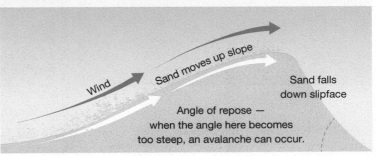

Wind

Sand moves up slope

Sand falls down slipface

Angle of repose — when the angle here becomes too steep, an avalanche can occur.

FIGURE 3 A series of crescent dunes in Egypt

FIGURE 4 Linear sand dunes in the Simpson Desert, Australia

Star dunes have 'arms' that radiate from a high central pyramid-shaped mound (figure 5). They form in regions that have winds blowing in many different directions and can become very tall rather than wide — some are up to 500 metres high.

Dome dunes are made up of fine sand without a steep side. These rounded structures tend to be only one or two metres high and are very rare (figure 6).

Parabolic dunes have a U shape and do not get very high (figure 7). They often occur in coastal deserts. The longer section follows the 'head' of the dune (the opposite process to the formation of crescent dunes) because vegetation has anchored them in place. The arms can be long — in one case, measured at 12 kilometres.

4.5.3 Playas and pans = another depositional landform

A desert basin may fill with water after heavy rains to form a shallow lake, but for the majority of the time the often salt-encrusted surface is hard and dry. Such expanses of land are known as playas, saltpans or hardpans. The flat terrains of pans and playas make them excellent race tracks and natural runways for aeroplanes and spacecraft. Ground-vehicle speed records are commonly established on Bonneville Speedway, a race track on the Great Salt Lake hardpan (figure 8). Space shuttles land on Rogers Lake playa at Edwards Air Force Base in California in the western United States.

FIGURE 5 Star dunes are found in many deserts including the Namib, the Grand Erg Oriental of the Sahara, and the south-east Badain Jaran Desert of China.

FIGURE 6 A dome dune in the Chihuahuan Desert, North America

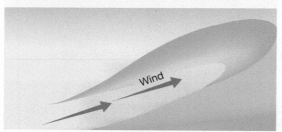

FIGURE 7 Formation of a parabolic dune

Wind

FIGURE 8 A driver lying in a streamlined racing car, Bonneville Salt Flats, Utah, the United States

4.5 Activities

To answer questions online and to receive **immediate feedback** and **sample responses** for every question, go to your learnON title at www.jacplus.com.au. *Note:* Question numbers may vary slightly.

Remember

1. List the agents of erosion and weathering in a desert. How does each process cause change in a desert?
2. Name two erosional and two depositional landforms in a desert.
3. Name the most common dune shapes that are formed in deserts.

Explain

4. Explain the difference between a mesa and a butte.
5. How does vegetation help to prevent erosion in a desert?
6. What wind conditions are needed to create a:
   (a) star dune
   (b) longitudinal dune
   (c) parallel dune?
7. Why do you think oases are such fertile places?
8. What do chimney rocks and arches have in common?

Discover

9. Locate all the desert *places* named in this subtopic. Use Google Maps to create your own map of these locations, and add some interesting facts and images of each location. Email a link to your completed map to your teacher.

Predict

10. Study the landforms labelled 1, 3 and 9 in figure 1. Sketch what each of these may look like in the future as erosion and weathering continue to occur.

Think

11. What do playa lakes and saltpans have in common?
12. Draw up a table like the one below.
    Continue to add the landforms shown in figure 1 to your table. Add examples of other desert landforms that you have found when researching this topic.

Name of landform	Picture of landform	Location	Type of erosion (wind or water)	Type of deposition (wind or water)
Butte				
Mesa				
Inselberg				

13. Work in small groups to create a model of a desert (using plasticine or playdoh, for example) that contains a number of desert forms and patterns. Use figure 1 as a guide. Show your completed model to the other groups, then provide and respond to constructive feedback.

4.6 What are the characteristics of Australia's deserts?

4.6.1 The location of Australia's deserts

Australia is the world's driest inhabited continent. Over 70 per cent of the country receives between 100 and 350 millimetres of rainfall annually, which makes most of Australia arid or semi-arid.

Australia's deserts are subtropical and are located mainly in central and western Australia, making up about 18 per cent of the country (see figure 1). They are hot deserts, which means they are areas of little rainfall and extreme temperatures — rainfall can be less than 250 millimetres per year and temperatures can rise to over 50 °C. The average humidity is between 10 and 20 per cent. The desert terrain is very diverse and can range from red sand dunes to the polished stones of the gibber plains – the term *gibber* comes from an Aboriginal language word for stone.

FIGURE 1 The location and distribution of Australia's deserts

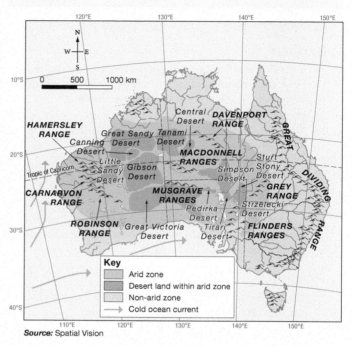

Source: Spatial Vision

Great Victoria Desert

The Great Victoria Desert, Australia's largest, covers 424 400 square kilometres. It is not a desert of dunes, but has some desert-adapted plants including marble gums, mulga and spinifex grass. Part of this desert has been named a Biosphere Reserve by UNESCO and is one of the largest arid zone biospheres in the world.

Great Sandy Desert

The Great Sandy Desert makes up 3.5 per cent of Australia. The red sands of this desert reach almost to the Western Australia coast, where they join with the white sand of Eighty Mile Beach south of Broome.

Simpson Desert

The Simpson Desert is in one of the driest areas of Australia, with rainfall of less than 125 millimetres per year. It is located near the geographical centre of Australia. Dunes (see figure 2) make up nearly three-quarters of the desert. Long parallel dunes (see figure 4 in subtopic 4.5) form in a north–north-west/south–south-east direction; some can be straight and unbroken for up to 300 kilometres and can be 40 metres high. The space between the dunes can vary from 100 metres to 1000 metres.

Strzelecki Desert

This desert is located within three states — far northern South Australia, south-west Queensland and western New South Wales. The dunes support vegetation such as sandhill wattle, needlebush and hard spinifex.

Tanami Desert

Located to the east of the Great Sandy Desert, this desert is mostly characterised by red sand plains with hills and ranges.

FIGURE 2 Sand dunes and vegetation in the Simpson Desert

FIGURE 3 Gibber landscape in the Sturt Stony Desert in South Australia

Little Sandy Desert

The Little Sandy Desert is located in Western Australia and borders three other deserts. Its landforms are similar to those in the Great Sandy Desert. It includes a vast salt lake called Lake Disappointment.

Sturt Stony Desert

The Sturt Stony Desert, located in north-eastern South Australia, is a harsh gibber desert covered in closely spaced glazed stones (figure 3). These are left behind when the wind blows away the loose sand between the dense covering of pebbles. The desert also contains some dunes and hills that are resistant to weathering.

FIGURE 4 Desert between Oodnadatta and William Creek, South Australia

Tirari Desert

This small desert covers almost 1600 square kilometres and is located in far northern South Australia, east of Lake Eyre. It contains many linear (parallel) dunes and salt lakes. Cooper Creek runs through the centre of the desert, as do many other **intermittent creeks**. Where there is enough water — usually in waterholes — river red gums and coolabah gums will grow. Tall, open shrubland also occurs in some areas.

Gibson Desert

The fifth largest in Australia, the Gibson Desert is located in Western Australia and borders three other deserts. It consists of sand plains and dunes plus some low, rocky ridges. Some small salt-water lakes are also present in the south-western part of the desert.

Pedirka Desert

The Pedirka Desert in South Australia is Australia's smallest desert, located north-east of Oodnadatta. The lines of parallel red dunes run north-east to south-west, and the space between the dunes can be up to one kilometre. Hamilton Creek is located in this desert and its banks are home to river red gums, coolabah, mulga and prickly wattle. Other vegetation includes satiny bluebush, weeping emubush and spiny saltbush. Common grasses include woollybutt, broad-leaf wanderrie, mulga grass and bandicoot grass.

4.6 Activities

To answer questions online and to receive **immediate feedback** and **sample responses** for every question, go to your learnON title at www.jacplus.com.au. *Note:* Question numbers may vary slightly.

Remember

1. Name the deserts bordered by:
   (a) the Gibson Desert
   (b) the Little Sandy Desert.
2. What is a gibber desert?
3. What percentage of Australia is arid or semi-arid?

Explain

4. Look at figure 1 showing the distribution of Australia's deserts. Where are they located in terms of the tropics?

Discover

5. Several plants are listed in the descriptions in this subtopic on Australia's deserts. Choose two different plant types (for example, a grass and a tree) and research how they are adapted to desert conditions.
6. Research the characteristics of the Biosphere Reserve declared by UNESCO that is located in the Great Victorian Desert.

Think

7. Use an atlas to find the locations of Brisbane, Geraldton and Exmouth. These *places* are located at the same latitude as many of Australia's deserts. Use the **BOM** weblink in the Resources tab to find the average temperature, rainfall and humidity of these *places*.
   (a) How do these characteristics compare with the temperature, rainfall and humidity in Australia's deserts?
   (b) How can you account for the differences?

---

learn on RESOURCES — ONLINE ONLY

 Explore more with this weblink: BOM

4.7 SkillBuilder: Calculating distance using scale

WHAT DOES IT MEAN TO CALCULATE DISTANCE USING SCALE?

Calculating distance using scale involves working out the actual distance from one place to another using a map. The scale on a map allows you to convert distance on a map or photograph to distance in the real world. A linear scale is the easiest to use.

Go online to access:
- a clear step-by-step explanation to help you master the skill
- a model of what you are aiming for
- a checklist of key aspects of the skill
- a series of questions to help you apply the skill and to check your understanding.

FIGURE 1 Calculating distance using a linear scale on a map

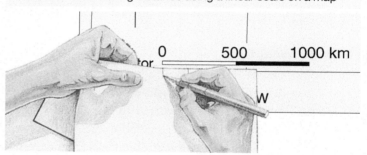

4.8 How did Lake Mungo become dry?

4.8.1 Where are Lake Mungo and the Willandra Lakes located?

Lake Mungo, in Mungo National Park, is just one of 13 ancient dry lake beds in a section of the Willandra Lakes Region World Heritage area in semi-arid New South Wales. There is no water there now, yet the lakes were once full of water and teeming with life, supporting Aboriginal peoples since the beginning of the Dreamings (more than 47 000 years by European estimates) — archaeological records show this continuous human presence. What happened to change this environment into the semi-arid landscape it is today?

The Willandra Lakes are located in far south-western New South Wales and the region is part of the Murray–Darling River Basin. Lake Mungo is 110 kilometres north-east of Mildura, Victoria. The lakes were originally fed by water from Willandra Creek (see figure 1), which was a branch of the Lachlan River. The average rainfall in this area is 325 millimetres per year, making it a semi-arid desert region.

4.8.2 How has Lake Mungo changed over time?

40 000 years ago

During the last ice age, huge amounts of water filled the shallow lake. At its fullest, Lake Mungo was 6–8 metres deep and covered 130 square kilometres (more than twice the area of Sydney Harbour). The lakes were rich with life, including water birds, freshwater mussels, yabbies and fish such as golden perch and Murray cod. Giant kangaroos, giant wombats, large emus and the buffalo-sized Zygomaturus

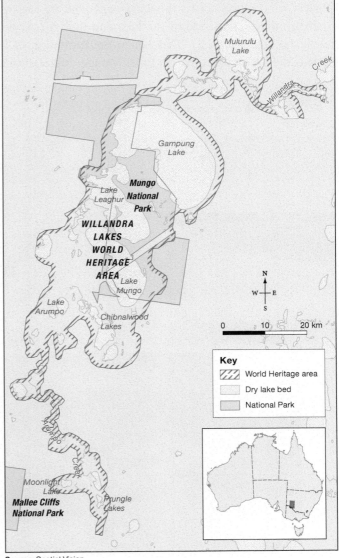

FIGURE 1 Location of the Willandra Lakes, including Lake Mungo

Source: Spatial Vision

— all now extinct — grazed around the water's edge. Remains of more than 55 species have been found in the area and identified — 40 of these are no longer found in the region, and 11 are extinct.

Aboriginal peoples lived here in large numbers — evidence for this has been found in more than 150 human fossils, including 'Mungo Lady' discovered in 1968 and 'Mungo Man' in 1974, both believed to be over 40 000 years old. The youngest fossil is 150 years old.

FIGURE 2 Traditional owners clean fossilised footprints at Lake Mungo.

30 000–19 000 years ago

A west wind blows across this landscape. During low-water years, red dust and clay were blown across the plains to the eastern side of the lake and they mixed with the sand dunes on the edge of the lake (formed when the lake was full). This began the formation of lunettes (crescent-shaped dunes) on the east side – called 'the Walls of China' in Lake Mungo. Vegetation covered the dunes, protecting them.

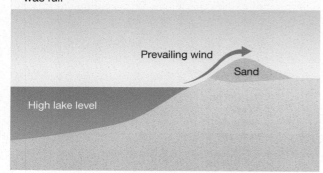

FIGURE 3 The process of erosion while Lake Mungo was full

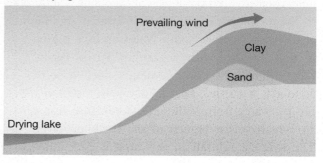

FIGURE 4 The process of erosion while Lake Mungo was drying out

FIGURE 5 The 'Walls of China' at Lake Mungo. The dry lake bed is covered by low bushes and grasses.

19 000 years ago

The lakes were full of deep, relatively fresh water for a period of 30 000 years — with cycles of wet and dry occurring — which came to an end 19 000 years ago when the climate became drier and warmer. Eventually, the water stopped flowing into the lake system and it dried out.

Present day

Today, the lake beds are flat plains covered by low saltbush and bluebush as well as grasses. Grazing cattle and sheep (now no longer allowed in the national park) and rabbits have caused erosion of the lunettes and sand dunes, exposing the human and animal fossils that have since been discovered.

4.8.3 World Heritage listing

The Willandra Lakes Region, which includes Lake Mungo, is listed as a World Heritage Area. This region is important because of its archaeology (human skeletons, tools, shell middens and animal bones make up the oldest evidence of burial places in the world) and geomorphology (ancient and undisturbed landforms and sediments).

4.8 Activities

To answer questions online and to receive **immediate feedback** and **sample responses** for every question, go to your learnON title at www.jacplus.com.au. *Note:* Question numbers may vary slightly.

Remember

1. What makes Lake Mungo and the Willandra Lakes region a semi-desert?
2. When did Lake Mungo dry up?
3. What is a lunette?

Explain

4. Describe how the lunettes formed over time.
5. Outline the evidence that shows that many Aboriginal peoples lived in this area.
6. What human activity caused the lunettes to erode? What did the erosion unearth?
7. Research what World Heritage listing means in terms of protecting this *place*. Why is it important?

Discover

8. Work in small groups to create an identification brochure with pictures and facts about these three extinct animals that once lived at Lake Mungo.
   • *Genyornis newtoni* (giant emu)
   • *Protemnodon goliah* (giant short-faced kangaroo)
   • *Zygomaturus trilobus* (Zygomaturus)

Think

9. Use the **Timetoast** weblink in the Resources tab, the information in this subtopic and images you find through online research to create your own colourful digital timeline of these changes that occurred at Lake Mungo.

RESOURCES — ONLINE ONLY

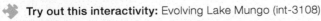

Try out this interactivity: Evolving Lake Mungo (int-3108)

Explore more with this weblink: Timetoast

4.9 How do people use deserts?

Access this subtopic at **www.jacplus.com.au**

4.10 Is Antarctica a desert?

4.10.1 Some facts about Antarctica

Like hot deserts, polar deserts are areas with annual precipitation of less than 250 millimetres, but they have a mean temperature during the warmest month of less than 10 °C. Polar deserts are found in both the Arctic and Antarctic regions of the world. Not only is Antarctica a desert — it is also the driest, coldest and windiest continent on Earth.

FIGURE 1 This cross-section, which shows the mountains below the ice, passes through some of the thickest parts of the Antarctic ice sheet.

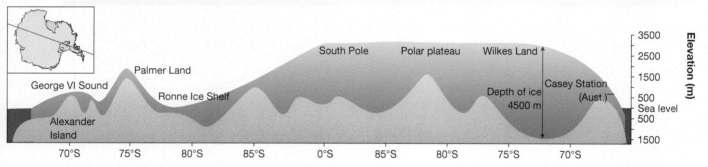

Australia is the driest inhabited continent on Earth. However, Antarctica is even drier. Much of Australia's interior receives less than 250 millimetres of precipitation per year. The interior of Antarctica receives less than 50 millimetres. The coastal areas receive the highest levels of precipitation, but this is still only about 200 millimetres.

Most of Antarctica is too cold for rainfall; the majority of the precipitation falls as snow. Some valleys in Antarctica have received no rain for two million years. It also snows very little in Antarctica, particularly in the interior.

In places, the ice sheet in Antarctica is 4.8 kilometres deep (see figure 1). Most of the ice that covers the continent has been there for thousands of years. In winter, as the surrounding oceans freeze, the area of Antarctica is almost double that in summer.

How dry is dry?

Covered in ice, Antarctica may seem like the wettest place in the world, but it's actually drier than the Sahara Desert. Despite this, Antarctica's ice holds 70 per cent of the world's fresh water supply.

FIGURE 2 A climograph for Mawson Base, Antarctica

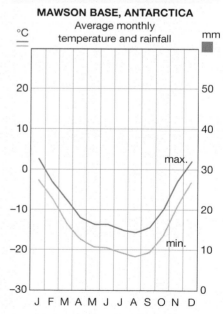

Note: Precipitation = zero

Most places in Antarctica receive no rain or snow at all. Very cold air does not have the capacity to hold enough water to create rain or snow. This means that Antarctica is the world's biggest desert. All drinking water in Antarctica is obtained by melting the ice. Unlike in hot deserts, there is little evaporation from Antarctica, so the relatively small amount of snow that does fall doesn't disappear. Instead it builds up over hundreds and thousands of years into enormously thick ice sheets.

How cold is cold?

On 29 July 1983, Russian scientists at Vostok Base, high on Antarctica's polar plateau, recorded a temperature of –89.2 °C, the lowest on record. During the coldest months (July to August), the average temperature at the South Pole is –60 °C. During the warmest months (December to January), it rises to –28 °C.

Why is Antarctica so cold?

There are three main reasons:

1. Antarctica's position on the globe means that the sun's rays strike the Earth's surface at a low angle, and therefore have a much larger area to heat than at other places on the planet.
2. Most of the sun's heat that does reach Antarctica is reflected back into space by the white ice that covers the continent. This also explains why you must always wear sunglasses or goggles in Antarctica.
3. Antarctica is surrounded by the cold waters of the Southern Ocean.

How windy is windy?

Australia's greatest polar explorer, Douglas Mawson, called Antarctica 'the home of the blizzard'. He should know. He lived in a wooden hut for two complete Antarctic winters, in the strongest winds ever recorded. Mawson's measurements revealed an average wind speed of over 70 kilometres per hour and gusts of over 300 kilometres per hour! The men in his expedition team always carried an ice axe with them to avoid being blown into the sea.

FIGURE 3 Sastrugi lie in the direction of the prevailing wind. They are as hard as rock.

Why is Antarctica so windy?

As the air over the polar plateau becomes colder, it becomes more dense. Finally gravity pulls it down off the plateau towards the Antarctic coast. This creates very strong winds, called **katabatic winds**, which can blow continually for weeks on end and carry small pellets of ice. These winds combined with the severe cold can be fatal; at –20 °C, exposed human flesh begins to freeze when the wind reaches only 14 kilometres per hour.

Katabatic winds also cause **blizzards**, which sweep up loose snow and blow it about ferociously. Such blizzards were the cause of death among many early Antarctic explorers.

The winds also shape the landscape, carving it into irregular shapes called **sastrugi** (see figure 3). These shapes range in height from 150 millimetres to two metres. Travelling across sastrugi is extremely difficult.

4.10 Activities

To answer questions online and to receive **immediate feedback** and **sample responses** for every question, go to your learnON title at www.jacplus.com.au. *Note:* Question numbers may vary slightly.

Remember

1. List three facts about Antarctica that you found the most surprising.
2. Why does Antarctica double in area every winter?
3. What is the coldest temperature ever recorded in Antarctica, and in which *place* did this occur?

Explain

4. Antarctica is sometimes described as the world's biggest desert. Why?
5. Describe and explain why Antarctica is so dry, cold and windy.
6. Examine the photograph in figure 3 and describe how this landscape has been formed. How does this *environment* pose a risk to people?

Discover

7. Use the **Antarctic weather** and **BOM** weblinks in the Resources tab to describe the weather conditions now at the South Pole. Compare these to the conditions where you live.
8. Use an atlas to measure the distance from Antarctica (coastline) to South America, Australia and South Africa.

Predict

9. What might happen to Antarctica if the ice shelves on top of the mountains were to melt? What *changes* might happen to sea levels around the world? Work with a partner to construct a concept map to record all your ideas.

Think

10. Use the information in table 1 to draw a climograph of McMurdo Station. How does it compare to Mawson Base (see figure 2)? Find climate data for the *place* where you live and draw another climograph for that location. Compare this to the two Antarctic climographs. Outline the similarities and differences and provide reasons for these.

TABLE 1 Climate data for the American McMurdo station in Antarctica: latitude 77.88°S, longitude 166.73°E

	Jan	Feb	Mar	Apr	May	Jun	Jul	Aug	Sep	Oct	Nov	Dec	Annual
Average daily temperature (°C)	−2.9	−9.5	−18.2	−20.7	−21.7	−23.0	−25.7	−26.1	−24.6	−18.9	−9.7	−3.4	Mean −16.9
Mean monthly rainfall (mm)	15	21.2	24.1	18.4	23.7	24.9	15.6	11.3	11.8	9.7	9.5	15.7	Total 202.5

learn on RESOURCES — ONLINE ONLY

🔗 **Explore more with these weblinks:** Antarctic weather, BOM

4.11 How do people use Antarctica?

4.11.1 Introduction

Antarctica and the seas that surround it contain valuable resources. Antarctica is also the temporary home of more than 4000 people in summer and 1000 in winter. Most are scientists and support staff. These people work in more than 66 research stations representing 30 different nations.

FIGURE 1 Esperanza, a permanent, all-year round Argentinian research base, Graham Land, Antarctica

4.11.2 Mining

There are great difficulties in looking for mineral deposits in rocks that lie beneath thousands of metres of ice. Therefore, most exploration has taken place in the ice-free areas of Antarctica. Scientists now believe there are deposits of many valuable minerals in Antarctica, including coal, iron ore, copper, lead and uranium, and traces of minerals such as gold and zinc. There are also mineral beds lying under the continent's Transantarctic Mountains, and large areas that may contain deposits of oil and gas (see figure 2).

Despite the presence of these valuable minerals, there are no operating mines in Antarctica. Given the conditions — the extreme cold, the rough seas and the wind — mining operations would be very difficult and potentially dangerous to the environment. Mining (other than for scientific purposes) is banned under the Antarctic Treaty. This is to prevent the possibility of polluting the environment (for example, through an oil spill or by digging a quarry).

The Antarctic Treaty

By the mid-1950s, Australia, New Zealand, the United Kingdom, France and Argentina were actively exploring Antarctica. These countries declared territorial claims over parts of Antarctica while others were fishing, whaling and conducting scientific research and mineral exploration in the region.

FIGURE 2 Potential sources of minerals in Antarctica

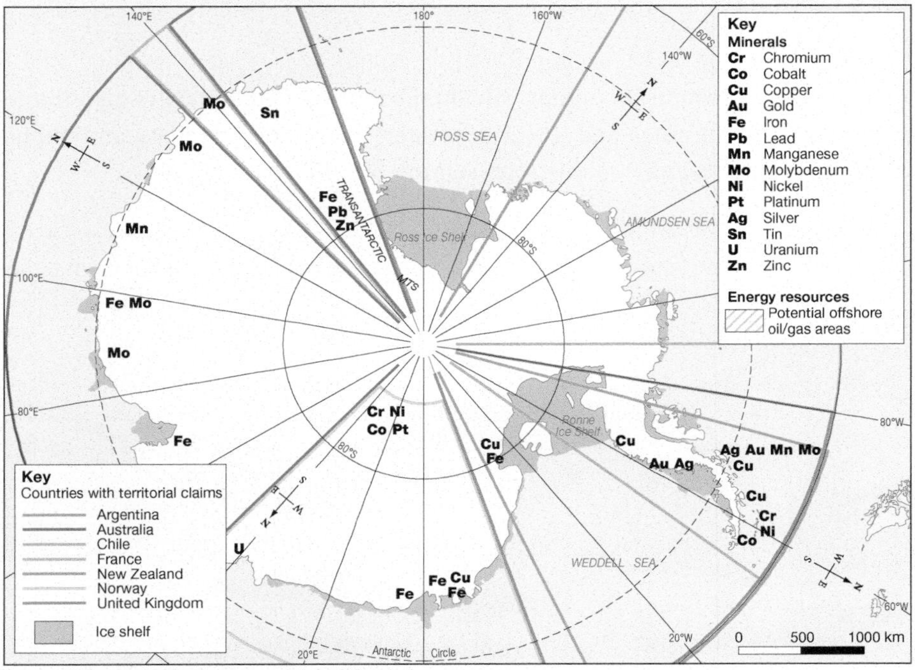
Key
Minerals
Cr Chromium
Co Cobalt
Cu Copper
Au Gold
Fe Iron
Pb Lead
Mn Manganese
Mo Molybdenum
Ni Nickel
Pt Platinum
Ag Silver
Sn Tin
U Uranium
Zn Zinc
Energy resources
Potential offshore oil/gas areas
Key
Countries with territorial claims
Argentina
Australia
Chile
France
New Zealand
Norway
United Kingdom
Ice shelf
ROSS SEA
AMUNDSEN SEA
WEDDELL SEA
Ross Ice Shelf
Ronne Ice Shelf
TRANSANTARCTIC MTS
Antarctic Circle
0 500 1000 km
140°E
180°
160°W
140°W
120°E
100°E
80°E
20°E
20°W
80°W
60°W
60°S
80°S

Source: © Geography Teachers' Association of Victoria, Inc.

People began to realise that this unique wilderness needed to be protected. In 1958, 12 countries agreed to preserve Antarctica. This led to an international agreement called the Antarctic Treaty, which came into force in 1961. The **treaty** covers the area south of 60 °S latitude. It has been signed by more than 52 countries who meet regularly to discuss issues affecting Antarctica. The treaty:

- prohibits military activity
- protects the Antarctic environment
- fosters scientific research
- recognises the need to protect Antarctica from uncontrolled destruction and interference by people.

4.11.3 Tourism

The number of tourists to Antarctica has increased significantly since the mid-1990s, with a peak of more than 45 000 in 2007–2008. However, more people will attend one game of AFL football in Melbourne than will visit Antarctica in one year. Given the scale (size) of Antarctica, tourist numbers are therefore still small.

Most tourists go to Antarctica on board cruise ships. There are opportunities for people to land on the ice. This often requires use of a Zodiac inflatable boat between ship and shore. There are no tourist facilities on Antarctica — people must return to the ships, for example, to sleep, eat and shower.

Sightseeing is the main activity for tourists. Other activities include kayaking, visiting research stations, walking and snowboarding. Other types of tourism include flights over the continent and flights that include landing on the ice.

Tourism can create problems, such as pollution from oil spills and disturbance to animal colonies. Therefore, the International Association of Antarctica Tour Operators has set up rules to control tourism. For example, no more than 100 passengers from a cruise ship may be landed at a location in Antarctica at any one time.

TABLE 1 Tourists to Antarctica

Year	Tourist numbers	Year	Tourist numbers
1996–97	7330	2006–07	29 823
1997–98	9604	2007–08	46 069
1998–99	10 013	2008–09	37 858
1999–2000	14 762	2009–10	36 975
2000–01	12 248	2010–11	33 824
2001–02	11 588	2011–12	26 509
2002–03	13 571	2012–13	34 354
2003–04	27 537	2013–14	37 405
2004–05	27 950	2014–15	36 702
2005–06	29 823		

Source: International Association of Antarctica Tour Operators

Bases on ice

Most of Antarctica's scientific bases are located on the coast so people and supplies can be brought in by boat or air (see figure 3). They are also situated on the two per cent of Antarctica not covered in ice, as bases built on ice tend to sink under their own weight. This is because the heat they generate can melt ice around and beneath them.

Some bases are inland. There is even a permanent scientific base at the South Pole: the American Amundsen–Scott Base. Australia operates three permanent bases in Antarctica — Casey, Mawson and Davis stations — plus one on Macquarie Island and five temporary summer bases.

In January 2008 an air link between Australia and Antarctica was officially opened. The Wilkins runway is a four-kilometre-long airstrip about 70 kilometres from Casey Station. Scientists can now get to Antarctica in a few hours from Australia rather than a few weeks on a ship. They can study the world's weather, climate, marine and land biology, glaciers, magnetics, geology and the ozone layer, as well as human physiology. Ice cores can provide a record of climate change over a long period of time.

FIGURE 3 Davis is the most southerly Australian Antarctic station.

4.11 Activities

To answer questions online and to receive **immediate feedback** and **sample responses** for every question, go to your learnON title at www.jacplus.com.au. *Note:* Question numbers may vary slightly.

Remember

1. List three ways in which the stations might have an impact on the Antarctic *environment*.
2. Why is there no mining in Antarctica? What problems would there be in extracting and transporting minerals from Antarctica?

Explain

3. Suggest the ideal location for a scientific base in Antarctica.
4. Why don't tourists visit Antarctica during winter?

Discover

5. Working in groups of four, use the **Life in Antarctica** weblink in the Resources tab to investigate life at the Australian Antarctic stations. Choose one station.
   (a) What facilities are there at the station?
   (b) Describe the work activities that take place.
   (c) What do you think it is like to live there?
6. Use a spreadsheet program to draw a line graph using the tourism data in table 1. Describe how the numbers have *changed* over time and provide one possible explanation for the drop in numbers since 2008.

Predict

7. Use the **Antarctic article** and **Biosecurity fears** weblinks in the Resources tab to find out more about how foreign seeds are invading Antarctica. Write a list of rules for a company that would remove this risk.

Think

8. Do you think countries should be able to own pieces of Antarctica? Write a two-minute speech outlining the reasons for your point of view. Debate this topic as a class.
9. Would you like to visit Antarctica? Why? Discuss as a class, listening carefully to the opinions of others.

  **learn** **on** RESOURCES — ONLINE ONLY

🔗 **Explore more with these weblinks:** Life in Antarctica, Antarctic article, Biosecurity fears

 my**World**Atlas   Deepen your understanding of this topic with related case studies and questions.
   ⊙ **Antarctica: human features**

4.12 Review

4.12.1 Review

The Review section contains a range of different questions and activities to help you revise and recall what you have learned, especially prior to a topic test.

4.12.2 Reflect

The Reflect section provides you with an opportunity to apply and extend your learning.

Access this subtopic at **www.jacplus.com.au**

TOPIC 5
Mountain landscapes

5.1 Overview

Numerous **videos** and **interactivities** are embedded just where you need them, at the point of learning, in your learnON title at www.jacplus.com.au. They will help you to learn the content and concepts covered in this topic.

5.1.1 Introduction

Mountains occupy 24 per cent of the Earth's landscape, and are characterised by many different landforms. The forces that form and shape mountains come from deep within the Earth, and have been shaping landscapes for millions of years. The Earth is a very active planet — every day, many volcanoes are erupting somewhere on the planet, and even more tremors are occurring.

Starter questions

1. What is the highest mountain you have visited? What did you do there?
2. Have you ever seen a volcano? Where was it?
3. Is there a mountain or mountain range you would like to visit? Why?
4. Have you ever felt an earthquake or earth tremor? Where were you when it happened? How did you react?

Landscape of the Dolomites, Italy

5.2 Where are the world's mountains?

5.2.1 The world's mountains and ranges

A mountain is a landform that rises high above the surrounding land. Most mountains have certain characteristics in common, although not all mountains have all these features. Many have steep sides and form a peak at the top, called a summit. Some mountains located close together have steep valleys between them known as gorges.

Mountains make up a quarter of the world's landscape. They are found on every continent and in three-quarters of all the world's countries. Only 46 countries have no mountains or high plateaus, and most of these are small island nations.

Some of the highest mountains are found beneath the sea. Some islands are actually mountain peaks emerging out of the water. Even though the world's highest peak (from sea level) is Mount Everest in the Himalayas (8850 metres high), Mauna Loa in Hawaii is actually higher when measured from its base on the ocean floor. Long chains or groups of mountains located close together are called a mountain range.

1 The Himalayas

Located in Asia, the Himalayas are the highest mountain range in the world. They extend from Bhutan and southern China in the east, through northern India, Nepal and Pakistan, and to Afghanistan in the west. The Himalayas is one of the youngest mountain ranges in the world and the name translates as 'land of snow'. The fourteen highest mountains in the world — all over 8000 metres above sea level — are all in the Himalayas.

2 The Alps

The Alps, located in south central Europe, are one of the largest and highest mountain ranges in the world. They extend 1200 kilometres from Austria and Slovenia in the east, through Italy, Switzerland, Liechtenstein and Germany, to France in the west. The highest mountain in the Alps is Mont Blanc in France, which is 4808 metres high.

3 The Andes

The Andes are located in South America, extending north to south along the western coast of the continent. The Andes is the second highest mountain range in the world, with many mountains over 6000 metres. At 7200 kilometres long, it is also the longest mountain range in the world.

4 The Rocky Mountains

The Rocky Mountains in western North America extend north–south from Canada to New Mexico, a distance of around 4800 kilometres. The highest peak is Mount Elbert, in Colorado, which is 4401 metres above sea level. The other large mountain range in North America is the Appalachian Mountains, which extends 2400 kilometres from Alabama in the south to Canada in the north.

FIGURE 1 The world's main mountains and mountain ranges

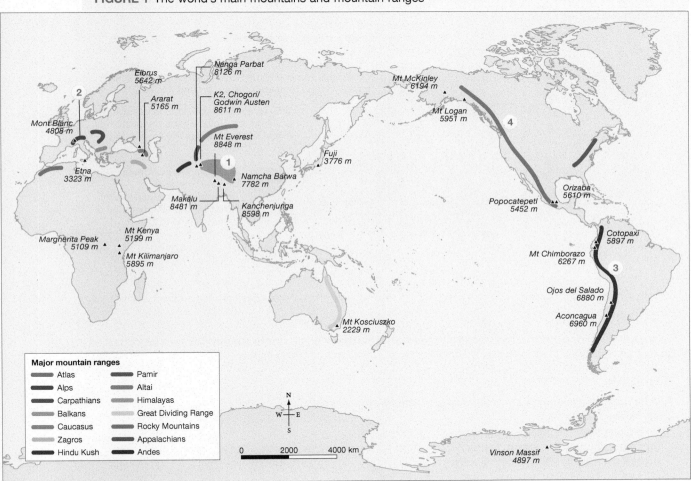

Source: Spatial Vision

5.2.2 Mountain climate and weather

It is usually colder at the top of a mountain than at the bottom, because air gets colder with **altitude**. Air becomes thinner and is less able to hold heat. For every 1000 metres you climb, the temperature drops by 6 °C.

FIGURE 2 Ecosystems change with altitude on mountains.

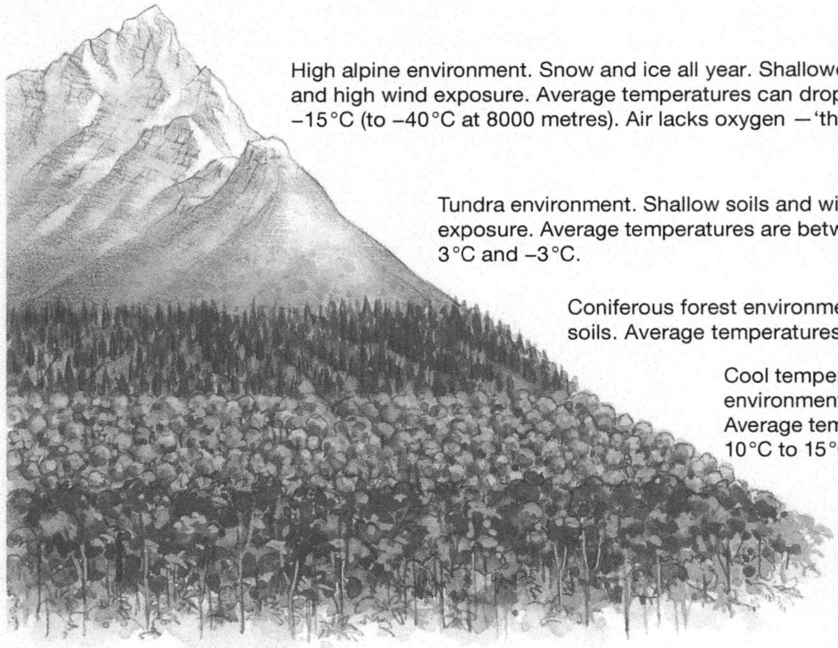

High alpine environment. Snow and ice all year. Shallowest soils and high wind exposure. Average temperatures can drop to –15 °C (to –40 °C at 8000 metres). Air lacks oxygen — 'thin air'.

Tundra environment. Shallow soils and wind exposure. Average temperatures are between 3 °C and –3 °C.

Coniferous forest environment. Shallow, slightly acidic soils. Average temperatures are around 5 °C to 9 °C.

Cool temperature deciduous forest environment. Soils with moderate humus. Average temperatures are around 10 °C to 15 °C.

Rainforest — evergreen forests with deep, relatively poor, leached soils. Base average temperature of around 20 °C to 25 °C.

5.2 Activities

To answer questions online and to receive **immediate feedback** and **sample responses** for every question, go to your learnON title at www.jacplus.com.au. *Note*: Question numbers may vary slightly.

Remember

1. What percentage of the Earth's surface is covered by mountains?
2. Name the:
   (a) highest mountain range in the world
   (b) longest mountain range in the world
   (c) highest mountain in Western Europe
   (d) second-highest mountain range in North America.
3. Refer to figure 2. How does vegetation *change* on a mountain?

Discover

4. Locate and name the highest mountain closest to where you live in your state or territory.
5. Name the mountain range located closest to where you live. Name two other mountain ranges in your state or territory. Where are they located in relation to your school or home?
6. Work in groups of 4 to 6 to investigate some of the following mountain ranges.
   • Antarctica — Antarctic Peninsula, Transantarctic Mountains
   • Africa — Atlas Mountains, Eastern African Highlands, Ethiopian Highlands
   • Asia — Hindu Kush, Himalayas, Taurus, Elburz, Japanese Mountains
   • Australia — MacDonnell Ranges, Great Dividing Range

- Europe — Pyrenees, Alps, Carpathians, Apennines, Urals, Balkan Mountains
- North America — Appalachians, Sierra Nevada, Rocky Mountains, Laurentians
- South America — Andes, Brazilian Highlands

Each student should choose a different range, and complete the following.

(a) Map the location of the range in its region.

(b) Describe the climate experienced throughout the range.

(c) Name and provide images of a selection of plants and animals found in the range.

Present your information in Google Maps.

Think

7. Imagine you are a mountaineer, climbing to the top of Mont Blanc.

(a) Research the type of clothing you need to wear for such a climb.

(b) When you begin your climb at 1500 metres, the weather is perfect; it is sunny and clear and the temperature is 8 °C. You climb 2200 metres before you set up camp. What is the elevation? What is the temperature at this elevation? The next day the weather holds, and you climb to the summit. How far did you climb to reach the top of the mountain? What is the temperature?

8. Refer to figure 1. Describe how the *scale* of the world's mountains varies across the continents.

5.3 How do people use mountains?

5.3.1 Mountain people and cultures

People have moved through and lived in mountain areas for centuries. But few people live in the world's highest mountain ranges, where it can be very cold and difficult to grow food and make a living. Thousands of people visit mountains, often in remote areas, for recreation and to see the spectacular scenery, plants and animals, historic and spiritual sites, and different cultures. Mountains are also vital for global water supply.

Around 12 per cent of the world's people live in mountain regions. About half of those live in the Andes, the Himalayas and the various African mountains.

Usually, population density is very low in these areas. One reason for this is that mountains are very difficult to cross, as they are often rugged and covered with forests and wild animals. They can also be hard to climb and may have ice, snow or glaciers that make travel dangerous.

As a result of these difficulties, mountains have long provided a safe place for **indigenous peoples** and **ethnic minorities**. People live as nomads, hunters, foragers, traders, small farmers, herders, loggers and miners.

FIGURE 1 The Longshen rice terraces in China show how a mountainside can be changed to grow food

5.3.2 Mountain landscapes in the Dreamings

There are many Aboriginal and Torres Strait Islander Dreaming Stories that are linked to mountain landscapes. These teachings from the Dreamings help explain the formation and importance of each landscape and landform.

Indigenous Australian peoples are guided by Elders who know the local Dreaming Stories and customs. Dreaming Stories are passed on through the generations and explain the origin of the world around them.

The Three Sisters in the Blue Mountains

There is a story, thought to be an Indigenous Creation Story, about the formation of the Three Sisters in the Blue Mountains in New South Wales (see figure 2). It tells of three sisters, Meehni, Wimlah and Gunnedoo, who lived in the Jamison Valley as members of the Gundungurra nations. These young women had fallen in love with three brothers from the Dharruk nation, yet tribal law forbade them to marry. The brothers were not happy with this law and so decided to use force to capture the three sisters, which caused a major battle.

As the lives of the three sisters were seriously in danger, a clever man from the Kedoombar took it upon himself to turn the three sisters into stone to protect them from any harm. He intended to reverse the spell when the battle was over, but the clever man himself was killed. As only he could reverse the spell and bring the sisters back to life, they remain in their rock formation.

FIGURE 2 The Three Sisters in the Blue Mountains

The Glasshouse Mountains

The Glasshouse Mountains in southeast Queensland are of great historical, cultural and geological significance (see figure 3). Their names — Beerwah, Tibrogargan, Coonowrin, Tunbubudla, Beerburrum, Ngungun, Tibberoowuccum and Coochin — reflect the culture of the Gubbi Gubbi people.

The story of these mountains goes something like this:

Tibrogargan was the father of all the nations. He and his wife, Beerwah, had many children, including Coonowrin, Tunbubudla, Miketeebumulgrai, Elimbah, Ngungun, Beerburrum and Coochin.

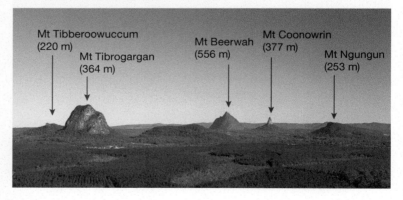

FIGURE 3 The Glasshouse Mountains

Mt Tibberoowuccum (220 m)

Mt Tibrogargan (364 m)

Mt Beerwah (556 m)

Mt Coonowrin (377 m)

Mt Ngungun (253 m)

One day, Tibrogargan was looking out to sea when he saw the sea rising in a great swell. He became worried for Beerwah, who was pregnant. He quickly told his eldest child, Coonowrin, to take his mother to the mountains. 'I'll get the other children together and will meet you there.'

But when Tibrogargan checked to see if Coonowrin had done as he had asked, he was angered to see that he was running off alone, like a coward, and had not fetched his mother. This made Tibrogargan angry. He chased Coonowrin and hit him so hard on the head with his nulla nulla (war club) that he dislocated his neck. It has been crooked ever since.

When the flood receded, the family went back to their lands. But when the others saw Coonowrin, they teased him about his crooked neck and how he came by it, making him ashamed of his cowardice. He asked his father to forgive him, but the law would not allow this. Tibrogargan cried many tears for the shame Coonowrin had brought upon them, and his tears formed a stream that went all the way to the sea. Beerwah and all Coonowrin's brothers and sisters cried too.

Coonowrin tried to explain that he had left his mother to fend for herself because she was so big. He did not know his mother was pregnant. Tibrogargan swore he would never look at his son again, and to this day he looks at the sea and not at Coonowrin, whose head is bowed and whose tears flow into the sea. As for Beerwah — she is still pregnant.

5.3.3 Sacred and special places

Mountain landscapes often have special meaning to certain groups of people. This might be because the location includes sacred sites or religious symbols; it might also be because people want to be close to nature or to feel spiritually inspired or renewed.

Mountaineers who take great risks, climbing alone or in small groups, often find a special meaning in mountain environments. They may hold deep spiritual, **cultural** and aesthetic (relating to beauty) values and ideas, and these will often inspire such people to care for and protect mountain environments.

The following list gives examples of mountains that are connected to various beliefs and religions.
- Hindus and Buddhists have beliefs about Mount Kailash in the Himalayas.
- Hindus in Bali, Indonesia, have a special connection with Mount Gunung Agung.
- Tibetan Buddhists revere Chomolungma (Mount Everest).
- The landscape of Demojong in the Himalayas is sacred to Tibetan Buddhists.
- Nanda Devi in the Himalayas is a sacred site for both Sikh and Hindu communities, and is a UNESCO World Heritage site.
- Mount Fuji, in Japan, is a place of spiritual and cultural symbolism to Japanese people.
- Saint Katherine Protectorate in South Sinai, Egypt, is in an area holy to Jews, Christians and Muslims.
- Jabal La'lam is a mountain that is sacred to the people of northern Morocco.

For the indigenous groups of the north-eastern American plains, the Sioux, or Dakota as they are sometimes referred to, and the indigenous Scandinavian people, the Sami, nature was recognized as sacred. The sacred places were not man-made temples or churches, but particularly spectacular or prominent features of the natural landscape. For the Sami these sacred places tended to be large rocks (called sieidi), the sides of lakes, rocky crevasses or caverns or mountaintops. These sacred mountains were somewhat isolated and had a jutting tall peak. A sacred mountain named Haldi, which rests among a group of mountains near Alta, and an 814-metre-tall conical sacred hill named Tunnsjøguden in central Norway are examples. In general, the word saivu is applied to sacred mountains in the south while the terms bassi, ailigas and haldi are used for sacred mountains by northern Sami. Similarly, mountaintops were also of spiritual importance to Sioux groups who lived in their regions; for instance, the sacred mountain Harney Peak in modern-day South Dakota.
Source: www.utexas.edu/courses/sami/diehtu/siida/religion/paralellism.htm

5.3.4 Skills to survive

It can be hard to make a living in mountain regions. People living in small, isolated mountain communities have learned to use the land and resources sustainably. Many practise shifting cultivation, migrate with grazing herds, and have terraced fields.

Some of the world's oldest rice terraces (see figure 1) are over 2000 years old. Rice and vegetables could be grown quite densely on the terraces. This enabled people to survive in a region with very steep slopes and high altitude.

On very high ranges, below the snowline is a treeless zone of alpine pastures that can be used in summer to graze animals. Elsewhere, in the valleys and foothills, agriculture often occurs, with fruit orchards and even vineyards on some sunny slopes.

Mountains supply 60 to 80 per cent of the world's fresh water. This is due to orographic rainfall (caused by warm, moist air rising and cooling when passing over high ground, such as a mountain; as the air cools, the water vapour condenses and falls as rain). Where precipitation falls as snow, water is stored in snowfields and glaciers. When these melt, they provide water to people when they need it most.

5.3 Activities

To answer questions online and to receive **immediate feedback** and **sample responses** for every question, go to your learnON title at www.jacplus.com.au. *Note*: Question numbers may vary slightly.

Remember

1. List the geographical characteristics of mountains that limit the number of people who live there.
2. What type of work and recreation can people undertake in mountain regions? Present this information in words or in a diagram.

Explain

3. Think of a mountain you have visited or seen. Do you feel inspired by mountain environments? How can spiritual or religious beliefs linked to mountain landscapes help in protecting them?
4. How has the natural mountain *environment* in figure 1 been *changed* by people? Sketch the photo and make notes to show the *changes*.
5. Imagine you work as a park ranger in the Blue Mountains or Glasshouse Mountains. How can the Dreamtime legends of the region help other people understand this *environment*?
6. Describe how different groups of people value mountainous *places*.

Discover

7. Use the internet to locate the Jamison Valley in the Blue Mountains. Describe its location.
8. Use the **Great Dividing Range legend** weblink and the **How the Hills Came to Be** weblink in the Resources tab to read two Dreamtime legends. How are each of these connected to mountain landscapes?
9. From this subtopic, choose one of the mountains linked to Hindu or Buddhist beliefs. Use the internet to find out details of this connection. Present your information as a print or electronic brochure.
10. Research where your water supply comes from. Which mountains, if any, are located near your water supply?

Think

11. Draw a consequence chart to show how and why mountains are important for water supply. Now add information to your chart about what might happen if this was reduced for some reason; for example, through climate *change*.

learn **on** RESOURCES — ONLINE ONLY

🔗 **Explore more with these weblinks:** Great Dividing Range legend, How the Hills Came to Be

5.4 What are the forces that form mountains?

5.4.1 Continental plates

Mountains and mountain ranges have formed over billions of years from tectonic activity; that is, movement in the Earth's crust. The Earth's surface is always changing — sometimes very slowly and sometimes dramatically.

The Earth's crust is cracked, and is made up of many individual moving pieces called continental plates, which fit together like a jigsaw puzzle. These plates float on the semi-molten rocks, or magma, of the Earth's mantle. Enormous heat from the Earth's core, combined with the cooler surface temperature, creates **convection currents** in the magma. These currents can move the plates by up to 15 centimetres per year. Plates beneath the oceans move more quickly than plates beneath the continents.

5.4.2 Continental drift

Scientific evidence shows that, about 225 million years ago, all the continents were joined.

FIGURE 1 World map of plates, volcanoes and hotspots

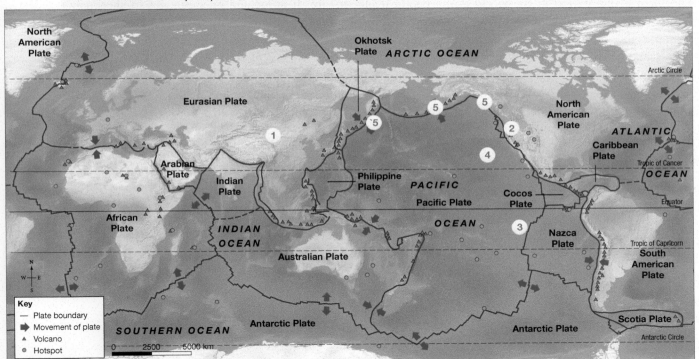

Source: Spatial Vision

① Convergent plates

When two continental plates of similar density collide, the pressure of the **converging plates** can push up land to form mountains. The Himalayas were formed by the collision of the Indian subcontinent and Asia. The European Alps were formed by the collision of Africa and Europe.

When an oceanic and a continental plate collide, they are different densities, and the thinner oceanic plate is subducted, meaning it is forced down into the mantle. Heat melts the plate and pressure forces the molten material back to the surface. This can produce volcanoes and mountain ranges. The Andes in South America were formed this way.

Subduction can also occur when two oceanic plates collide. This forms a line of volcanic islands in the ocean about 70–100 kilometres past the subduction line. The islands of Japan have been formed in this way. Deep oceanic trenches are also formed when this occurs. The Mariana Trench in the Pacific Ocean is

2519 kilometres long and 71 kilometres wide, and is the deepest point on Earth — 10.911 kilometres deep. If you could put Mount Everest on the ocean floor in the Mariana Trench, its summit would lie 1.6 kilometres below the ocean surface.

2 Lateral plate slippage

Convection currents can sometimes cause plates to slide, or slip, past one another, forming **fault** lines. The San Andreas Fault, in California in the western United States, is an example of this.

3 Divergent plates

In some areas, plates are moving apart, or diverging, from each other (for example, the Pacific Plate and Nazca Plate). As the **divergent plates** separate, magma can rise up into the opening, forming new land. Underwater volcanoes and islands are formed in this way.

4 Hotspots

There are places where volcanic eruptions occur away from plate boundaries. This occurs when there is a weakness in the oceanic plate, allowing magma to be forced to the surface, forming a volcano. As the plate drifts over the hotspot, a line of volcanoes is formed.

5 The Pacific Ring of Fire

The most active region in the world is the Pacific Ring of Fire. It is located on the edges of the Pacific Ocean and is shaped like a horseshoe. The Ring of Fire is a result of the movement of tectonic plates. For example, the Nazca and Cocos plates are being subducted beneath the South American Plate, while the Pacific and Juan de Fuca plates are being subducted beneath the North American Plate. The Pacific Plate is being subducted under the North American Plate on its east and north sides, and under the Philippine and Australian plates on its west side. The Ring of Fire is an almost continuous line of volcanoes and earthquakes. Most of the world's earthquakes occur here, and 75 per cent of the world's volcanoes are located along the edge of the Pacific Plate.

FIGURE 2 The Earth's core is very hot, while its surface is quite cool. This causes hot material within the Earth to rise until it reaches the surface, where it moves sideways, cools, and then sinks.

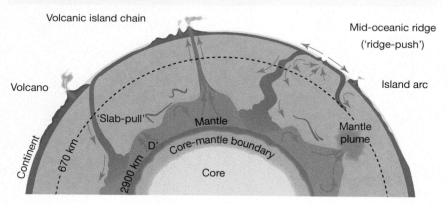

5.4 Activities

To answer questions online and to receive **immediate feedback** and **sample responses** for every question, go to your learnON title at www.jacplus.com.au. *Note*: Question numbers may vary slightly.

Remember

1. Look at figure 2. How do convection currents help explain plate tectonics?

Explain

2. Refer to figure 1. Name three *places* where plates are converging. What mountain ranges, if any, are located in these *places*?
3. Explain, in your own words, the meaning of subduction when referring to plate movements.
4. Name two locations where plates are moving apart. What is happening to the sea floor in these *places*?
5. Describe the distribution of volcanoes shown in figure 1. What does this distribution have in common with the location of plate boundaries?

Discover

6. Use different coloured strips of plasticine to make models showing how a collision of continental and oceanic plates differs from a collision of two continental plates. Have a go at explaining this to a member of your family.

Predict

7. Draw a sketch to show what you think the world's continents will look like many years into the future based on the way continents move and *change*. Justify your decisions.

Think

8. True or false?
   (a) The world's volcanoes are randomly scattered over the Earth's surface.
   (b) Most of the world's volcanoes are concentrated along the edges of certain continents.
   (c) Island chains are closely linked with the location of volcanoes.
   (d) There is a weak link between the distribution of volcanoes and the location of continental plates.
   Use these statements to write a summary paragraph, remembering to rewrite the false statements to make them true.

 RESOURCES — ONLINE ONLY

 **Try out this interactivity:** Mountain builders (int-3109)

my **World** Atlas   **Deepen your understanding of this topic with related case studies and questions.**
                       Active Earth

5.5 How do different types of mountains form?

5.5.1 Types of mountains

The different movements and interactions of the **lithosphere** plates result in many different mountain landforms. Mountains can be classified into five different types, based on what they look like and how they were formed. These are fold, fault-block, dome, plateau and volcanic mountains. (Volcanic mountains are discussed in subtopic 5.8.)

5.5.2 Fold mountains

The most common type of mountain, and the world's largest mountain ranges, are fold mountains. The process of folding occurs when two continental plates collide, and rocks in the Earth's crust buckle, fold and lift up. The upturned folds are called anticlines, and the downturned folds are synclines (see figure 2). These mountains usually have pointed peaks.

FIGURE 1 Selected world mountains

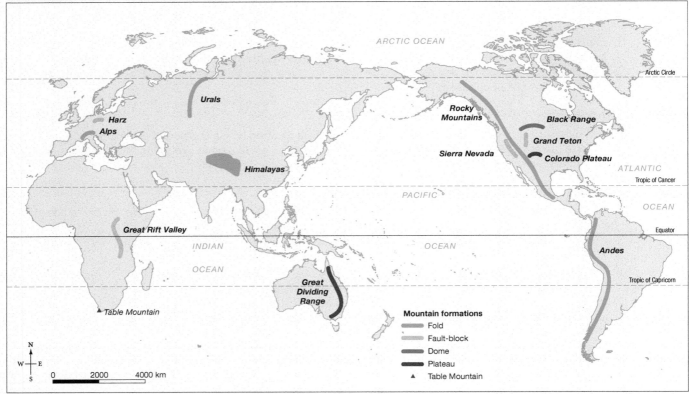

ARCTIC OCEAN

Arctic Circle

Urals

Harz
Alps

Himalayas

Rocky
Mountains

Black Range

Grand Teton

Sierra Nevada

Colorado Plateau

ATLANTIC

Tropic of Cancer

OCEAN

PACIFIC

Great Rift Valley

INDIAN

OCEAN

OCEAN

Equator

Andes

Great
Dividing
Range

Tropic of Capricorn

Table Mountain

Mountain formations
— Fold
— Fault-block
— Dome
— Plateau
▲ Table Mountain

N
W—E
S

0   2000   4000 km

Source: Spatial Vision

Use the **Anticline and syncline** weblink in the Resources tab to see the formation of fold mountains.

Examples of fold mountains include:
- the Himalayas in Asia
- the Alps in Europe
- the Andes in South America
- the Rocky Mountains in North America
- the Urals in Russia.

FIGURE 2 The formation of fold mountains

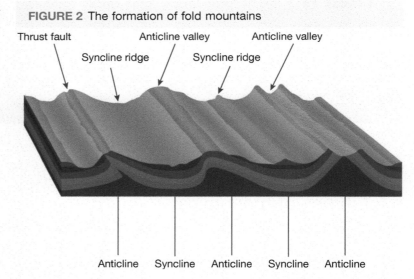

Thrust fault   Anticline valley   Anticline valley

Syncline ridge   Syncline ridge

Anticline   Syncline   Anticline   Syncline   Anticline

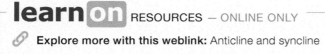

RESOURCES — ONLINE ONLY

🔗 **Explore more with this weblink:** Anticline and syncline

5.5.3 Fault-block mountains

Fault-block mountains form when faults (or cracks) in the Earth's crust force some parts of rock up and other parts to collapse down. Instead of folding, the crust fractures (pulls apart) and breaks into blocks. The exposed parts then begin to erode and shape mountains and valleys (see figure 3).

Fault-block mountains usually have a steep front side and then a sloping back. The Sierra Nevada and Grand Tetons in North America, the Great Rift Valley in Africa, and the Harz Mountains in Germany are examples of fault-block mountains. Another name for the uplifted blocks is *horst*, and the collapsed blocks are *graben*.

5.5.4 Dome mountains

Dome mountains are named after their shape, and are formed when molten magma in the Earth's crust pushes its way towards the surface. The magma cools before it can erupt, and it then becomes very hard. The rock layers over the hardened magma are warped upwards to form the dome. Over time, these erode, leaving behind the hard granite rock underneath (see figure 5).

Ben Nevis in Scotland is an example of a dome mountain (see figure 6).

FIGURE 3 The formation of fault-block mountains

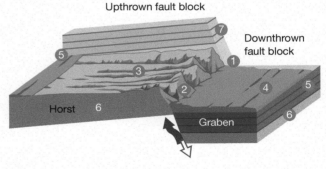

1. Fault zone
2. Steep eastern face
3. Gentle western slope
4. Valley floors filled with sediments of cobbles, gravel and sand
5. Sedimentary rock layers
6. Bedrock
7. Sedimentary rock layers (5) now worn away.

FIGURE 4 A cliff overlooking the Great Rift Valley in northern Tanzania, Africa. These are examples of fault-block mountains.

FIGURE 5 Very hot magma pushes towards the surface to form dome mountains.

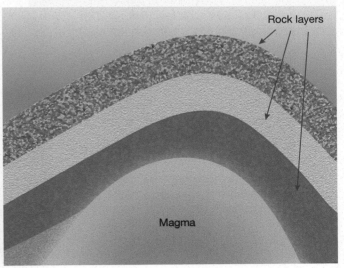

FIGURE 6 Ben Nevis in Scotland is a famous example of a dome mountain.

5.5.5 The Himalayas

How were the Himalayas formed?

Before the theory of tectonic plate movement, scientists were puzzled by findings of fossilised remains of ancient sea creatures near the Himalayan peaks. Surely these huge mountains could not once have been under water?

Since understanding plate movements, the mystery has been solved. About 220 million years ago, India was part of the ancient supercontinent we call **Pangaea**. When Pangaea broke apart, India began to move northwards at a rate of about 15 centimetres per year. About 200 million years ago, India was an island separated from the Asian continent by a huge ocean.

When the plate carrying India collided with Asia 40 to 50 million years ago, the oceanic crust (carrying fossilised sea creatures) slowly crumpled and was uplifted, forming the high mountains we know today. It also caused the uplift of the Tibetan Plateau to its current position. The Bay of Bengal was also formed at this time.

The Himalayas were therefore formed when India crashed into Asia and pushed up the tallest mountain range on the continents.

The Himalayas are known as young mountains, because they are still forming. The Indian and Australian plates are still moving northwards at about 45 millimetres each year, making this boundary very active. It is predicted that over the next 10 million years it will travel more than 180 kilometres into Tibet and that the Himalayan mountains will increase in height by about five millimetres each year. Old mountains are those that have stopped growing and are being worn down by the process of erosion.

FIGURE 7 The movement of the Indian landmass to its current location

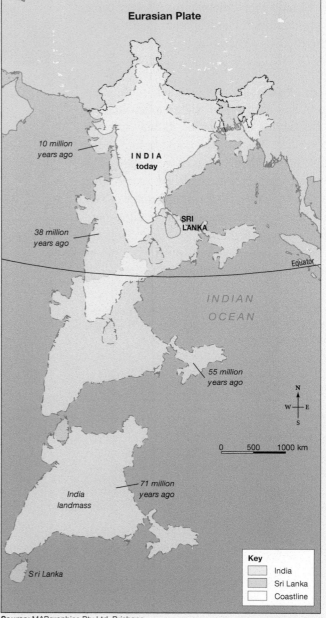

Source: MAPgraphics Pty Ltd, Brisbane

5.5.6 The Sierra Nevada Range, United States

CASE STUDY

Formation of the Sierra Nevada Range

The Sierra Nevada Range began to rise about five million years ago. As the western part of the block tilted up, the eastern part dropped down. As a result there is a long, gentle slope towards the west and a steep slope to the east.

FIGURE 9 Yosemite Valley in the Sierra Nevada mountains

FIGURE 8 The Sierra Nevada Range was formed by fault-block tilting.

5.5.7 Plateau mountains

Plateaus are high areas of land that are large and flat. They have been pushed above sea level by tectonic forces or have been formed by layers of lava. Over billions of years, streams and rivers cause erosion, leaving mountains standing between valleys. Plateau mountains are sometimes known as erosion mountains.

Examples of plateau mountains include Table Mountain in South Africa (see figure 10), the Colorado Plateau (see figure 11) in the United States and parts of the Great Dividing Range in Australia.

FIGURE 10 The plateau of Table Mountain towers over the city of Cape Town in South Africa.

FIGURE 11 The Colorado Plateau in the United States was raised as a single block by tectonic forces. As it was uplifted, streams and rivers cut deep channels into the rock, forming the features of the Grand Canyon.

5.5 Activities

To answer questions online and to receive **immediate feedback** and **sample responses** for every question, go to your learnON title at www.jacplus.com.au. *Note:* Question numbers may vary slightly.

Remember

1. List one example of fold, fault, dome and plateau mountains. Where is each located?
2. State whether the following statements are true or false.
   (a) Fold mountains are the most common type of mountain in the world.
   (b) The Sierra Nevada Range was formed by the eastern part of a fault-block tilting up.

Explain

3. Use the **Fold mountains** weblink in the Resources tab to explain the formation of fold mountains.
4. Use the video to explain the formation of fault-block mountains.
5. How does the shape of each of the mountains shown in this subtopic provide clues as to how they were formed? How have the effects of erosion *changed* these mountains?
6. Use figures 5 and 6 to explain the formation of dome mountains.

Discover

7. Sketch figure 6 and annotate it to show where erosion has taken place.
   Label *places* that have hard and weak rocks.
8. Draw a sketch of figure 11, noting the plateau and areas of erosion and weathering.
9. Use an atlas to locate the Sierra Nevada Range. Describe where it is. Name two national parks in this mountain range. Choose one, and investigate some of its geographical characteristics. Present this as a PowerPoint, Keynote or Prezi presentation.

Predict

10. Refer to the case study in this subtopic which describes the formation of the Himalayas.
    (a) Provide an explanation for why scientists found ancient sea fossils on top of the Himalayas.
    (b) Describe how the Himalayas were formed. How long did it take for the plate carrying India to crash into Asia? Explain why these mountains are described as 'young' mountains.
    (c) Based on the movements occurring, predict what might happen to the Himalayas in the future.

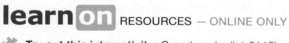 RESOURCES — ONLINE ONLY

 Try out this interactivity: Grand peaks (int-3110)

 Explore more with this weblink: Fold mountains

5.6 How do earthquakes and tsunamis occur?

5.6.1 Earthquakes and tsunamis

Earthquakes and tsunamis are frightening events and they often strike with little or no warning. An earthquake can shake the ground so violently that buildings and other structures collapse, crushing people to death. If an earthquake occurs at sea, it may cause a tsunami, which produces waves of water that move to the coast and further inland, sometimes with devastating effects.

5.6.2 Earthquakes

Earthquakes occur every day somewhere on the planet, usually on or near the boundaries of tectonic plates. The map in figure 1 subtopic 5.4 shows a strong relationship between the location of plate boundaries and the occurrence of earthquakes. Weaknesses and cracks in the Earth's crust near these plate boundaries are called faults. An earthquake is usually a sudden movement of the layers of rock at these faults.

The point where this earthquake movement begins is called the **focus** (see figure 1). Earthquakes can occur near the surface or up to 700 kilometres below. The shallower the focus, the more powerful the earthquake will be. Energy travels quickly from the focus point in powerful **seismic waves**, radiating out like ripples in a pond. The seismic waves decrease in strength as they travel away from the epicentre. The strength of an earthquake is measured on the Richter scale.

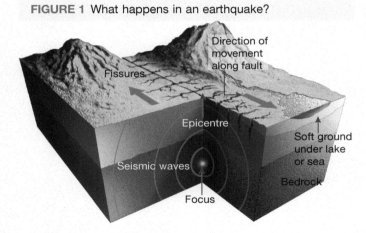

FIGURE 1 What happens in an earthquake?

The energy released at the focus can be immense, and it travels in seismic waves through the mantle and crust of the Earth. Primary waves, or P-waves, are the first waves to arrive, and are felt as a sudden jolt. Depending on the type of rock or water in which they are moving, these waves travel at speeds of up to 30 000 kilometres an hour.

Secondary waves, or S-waves, arrive a few seconds later and travel at about half the speed of P-waves. These waves cause more sustained up-and-down movement.

Surface waves radiate out from the epicentre and arrive after the main P-waves and S-waves. These move the ground either from side-to-side, like a snake moving, or in a circular movement.

Even very strong buildings can collapse with these stresses. The energy that travels in waves across the Earth's surface can destroy buildings many kilometres away from the epicentre.

Measuring earthquakes

Earthquakes are measured according to their magnitude (size) and intensity. Magnitude is measured on the Richter scale, which shows the amount of energy released by an earthquake. The scale is open-ended as there is no upper limit to the amount of energy an earthquake might release. An increase of one in the scale is 10 times greater than the previous level. For example, energy released at the magnitude of 7.0 is 10 times greater than the energy released at 6.0.

Earthquake intensity is measured on the Modified Mercalli scale, and indicates the amount of damage caused. Intensity depends on the nature of buildings, time of day and other factors.

5.6.3 Nepal earthquake, 2015

CASE STUDY

What caused the 2015 Nepal earthquake?

On 25 April 2015, a 7.8-magnitude earthquake struck Nepal at around midday. The epicentre of this earthquake was quite shallow — only 15 kilometres below the Earth's surface. It occurred approximately 80 kilometres to the north-west of Kathmandu, Nepal's capital.

At this location, the Indian Plate to the south is subducting under the Eurasian Plate to the north (see figure 1 in subtopic 5.4). This is occurring at a rate of approximately 45 millimetres per year and is causing the uplift of the Himalayas (see the case study in section 5.5.5).

During the Nepal earthquake event, nearly 9000 people were killed and nearly 18 000 were injured (see the case study in subtopic 5.7).

Figure 3 shows that the earthquake released a large amount of energy and caused large slips of up to four metres of the Earth's surface. There were severe aftershocks immediately after the main earthquake and the aftershocks continued for many weeks — up to 100 in total. The shaking from this earthquake was felt in China, India, Bhutan and much of western Bangladesh.

On 12 May 2015, a huge aftershock with a magnitude of 7.3 occurred near the Chinese border with Nepal (between Kathmandu and Mount Everest). More than 160 people died and more than 2500 were injured as a result of this aftershock.

FIGURE 2 The shake intensity and the tectonic plate boundary involved in the Nepal earthquake

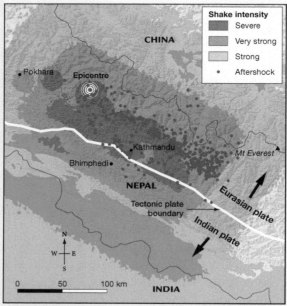

Source: USGS

FIGURE 3 Magnitudes of earthquake and aftershocks in Nepal, 2015

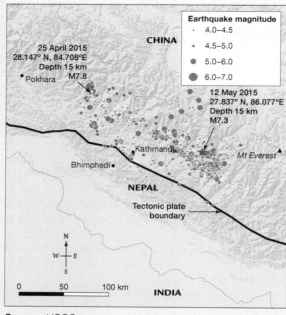

Source: USGS

5.6.4 Christchurch earthquake, New Zealand, 2011

What caused the 2011 Christchurch earthquake?

A 6.3-magnitude earthquake struck Christchurch, New Zealand, on 22 February 2011. The city was badly damaged, 185 people were killed and several thousand were injured. The earthquake epicentre was 10 kilometres south-east of Christchurch's central business district, and was quite shallow — only five kilometres deep. The earthquake is considered to be an aftershock of an earthquake that occurred five months earlier in September 2010.

New Zealand is located between two huge moving plates — the Australian Plate and the Pacific Plate — and it experiences thousands of earthquakes every year. Most are very small, but some have caused a lot of damage. These movements continue to shape and form New Zealand and its dramatic mountain landscapes.

FIGURE 4 Location of the Christchurch earthquake in New Zealand

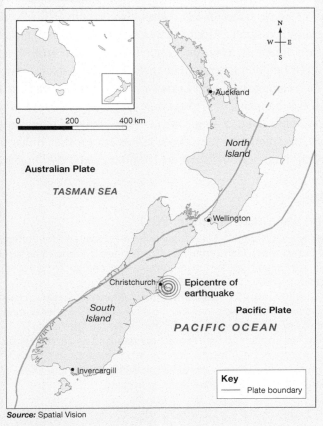

Source: Spatial Vision

FIGURE 5 Earthquake damage in Christchurch

5.6.5 Tsunamis

A tsunami is a large ocean wave that is caused by sudden motion on the ocean floor. The sudden motion could be caused by an earthquake, a volcanic eruption or an underwater landslide. About 90 per cent of tsunamis occur in the Pacific Ocean, and most are caused by earthquakes that are over 6.0 on the Richter scale (see figure 6).

A tsunami at sea will be almost undetectable to ships or boats. The reasons for this are that the waves travel extremely fast in the deep ocean (about 970 kilometres per hour — as fast as a large jet) and the wavelength is about 30 kilometres, yet the wave height is only one metre.

When tsunamis reach the continental slope, several things happen. The wave slows down and, as it does, the wave height increases and the wavelength decreases; in other words, the waves get higher and closer together. Sometimes, the sea may recede quickly, very far from shore, as though the tide has suddenly gone out. If this happens, the best course of action is to head to higher ground as quickly as possible.

A tsunami is not a single wave. There may be between five and 20 waves altogether. Sometimes the first waves are small and they become larger; at other times there is no apparent pattern. Tsunami waves will arrive at fixed periods between 10 minutes and two hours.

CASE STUDY

The Japanese tsunami, 2011

The region of Japan is seismically active because four plates meet there: the Eurasian, Philippine, Pacific and Okhotsk. Many landforms in this region are influenced by the collision of oceanic plates. Chains of volcanic islands called island arcs are formed, and an ocean trench is located parallel to the island arc (see figure 1 in subtopic 5.4).

On 11 March 2011, an 8.9-magnitude earthquake struck near the coast of Japan. The earthquake was caused by movement between the Pacific Plate and the North American Plate. It occurred about 27 kilometres below the Earth's surface along the Japan Trench, where the Pacific Plate moves westwards at about eight centimetres each year. The sudden upward movement released an enormous amount of energy and caused huge displacement of the sea water, causing the tsunami. When the tsunami reached the Japanese coast, waves more than six metres high moved huge amounts of water inland. Strong aftershocks were felt for a number of days. Nearly 16 500 people were killed and 4800 were reported missing.

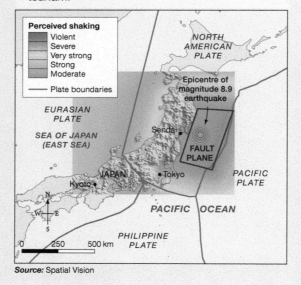

FIGURE 6 The location and magnitude of the earthquake that caused the Japanese tsunami

Source: Spatial Vision

FIGURE 7 This map shows the ground motion and shaking intensity from the earthquake across Japan.

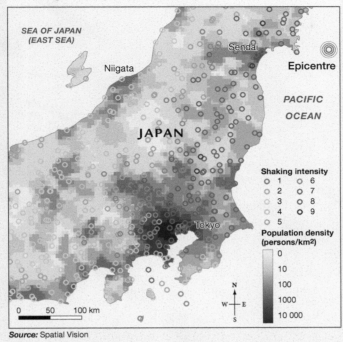

SEA OF JAPAN
(EAST SEA)

Niigata

Sendai

Epicentre

PACIFIC
OCEAN

JAPAN

Tokyo

Shaking intensity
1   6
2   7
3   8
4   9
5

Population density
(persons/km²)

0
10
100
1000
10 000

N
W—E
S

0   50   100 km

Source: Spatial Vision

FIGURE 8 The tsunami caused by the 8.9-magnitude earthquake in March 2011 swept over the coastline at Sukuiso and inland, carrying debris with it.

FIGURE 9 An earthquake and subsequent tsunami in the Indian Ocean in 2004 occurred along the boundary between tectonic plates.

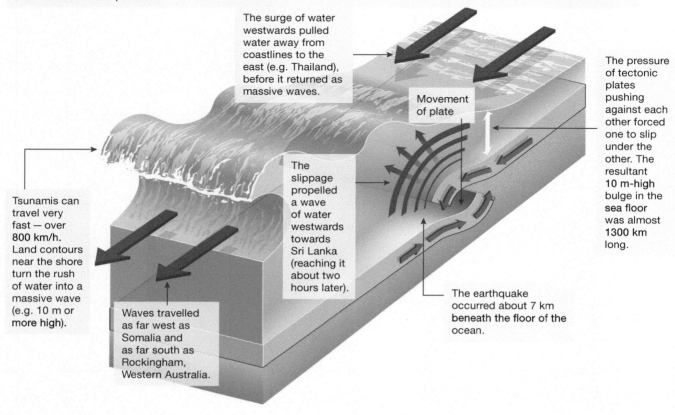

The surge of water westwards pulled water away from coastlines to the east (e.g. Thailand), before it returned as massive waves.

The pressure of tectonic plates pushing against each other forced one to slip under the other. The resultant 10 m-high bulge in the sea floor was almost 1300 km long.

Movement of plate

Tsunamis can travel very fast — over 800 km/h. Land contours near the shore turn the rush of water into a massive wave (e.g. 10 m or more high).

The slippage propelled a wave of water westwards towards Sri Lanka (reaching it about two hours later).

Waves travelled as far west as Somalia and as far south as Rockingham, Western Australia.

The earthquake occurred about 7 km beneath the floor of the ocean.

5.6 Activities

To answer questions online and to receive **immediate feedback** and **sample responses** for every question, go to your learnON title at www.jacplus.com.au. *Note*: Question numbers may vary slightly.

Remember

1. What is the focus and epicentre of an earthquake?
2. How does an earthquake occur?
3. What does the Richter scale measure? How much more powerful is the magnitude of an earthquake at 7.0 than at 5.0?

Explain

4. Use the **P- and S-waves** weblink in the Resources tab to watch an animation of these waves. What is the difference between the waves? How fast do they travel? How is damage caused by the waves?
5. Study figure 9. Use your own words to explain how a tsunami occurs.
6. Study figure 2.
   (a) In which direction is the Indian Plate moving? Is it moving under or over the Eurasian Plate?
   (b) Describe the location of the highest intensity shaking. How close was it to the epicentre? To the tectonic plate boundary?
7. Study figure 3. Are the following statements true or false? If they are false, rewrite them to make them true.
   (a) The earthquake and aftershocks were between 4.0 and 6.0 in magnitude.
   (b) The furthest earthquake and aftershocks were 100 km apart.
   (c) The earthquake on 12 May was the same intensity as the earthquake on 25 April.
   (d) Most of the aftershocks were felt to the east of the main earthquake on 25 April.

8. How does this earthquake event support the idea that the Himalayas are a young mountain range that is still forming?

Discover

9. Use an atlas or Google Earth to find the location of Lituya Bay. Draw a map to show the location. Use the **World's biggest tsunami** weblink in the Resources tab to listen to eyewitness accounts of the event. How does this help give you a sense of the *scale* of this event?
10. Study figure 6. Describe where the most violent shaking occurred as a result of the earthquake. How many plates meet in this region? What impact does this have?
11. Study figure 7.
    (a) Where in Japan was the greatest intensity felt?
    (b) What is the population density for Sendai, Tokyo and Niigata? How would this increase the impact of the earthquake?

Think

12. Study the photo of the Japanese tsunami in figure 8.
    (a) Imagine you are a radio news reporter. Describe what you see and what might be happening to people in the area.
    (b) Imagine you were a Sendai resident. Describe what you would have done to take care of yourself.
13. Why would most Australians not know what to do if an earthquake occurred?

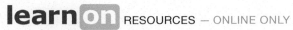 RESOURCES — ONLINE ONLY

 Try out this interactivity: Anatomy of a tsunami (int-3111)

 Explore more with these weblinks: P- and S-waves, World's biggest tsunami

---

 Deepen your understanding of this topic with related case studies and questions.
  ❯ Haiti earthquake
  ❯ Banda Aceh tsunami

5.7 What are the impacts of earthquakes and tsunamis?

Access this subtopic at **www.jacplus.com.au**

5.8 How are volcanic mountains formed?

5.8.1 How are volcanoes formed?

A volcano is a cone-shaped hill or mountain formed when molten magma in the Earth's mantle is forced through an opening or vent in the lithosphere. Almost all active volcanoes occur at or near plate boundaries. Some occur where two plates converge, and others occur where the plates are pulling apart, or diverging (see figure 1). There is another group of volcanoes that are formed when plates move over hotspots.

Subduction zones

Some volcanoes are formed when an oceanic plate is pulled underneath a continental plate (see subtopic 5.4). As the crust is forced down, it heats up and becomes magma. It can then rise to the Earth's surface through a magma chamber.

Volcanoes in rift zones

The longest mountain range in the world is underwater between the African and American continents, and is 56 000 kilometres long. It is called the Mid-Atlantic Ridge, and it is made up of many volcanic mountains. The volcanoes are formed where two plates move away from each other in **rift zones**. The molten lava rises to the surface in the space between the plates, and the largest volcanoes appear above the water as islands. Examples of rift islands are Iceland, the Azores, Ascension Island, Gough Island and Bouvet Island. The rifting, or spreading apart, can occur on land or on the seabed.

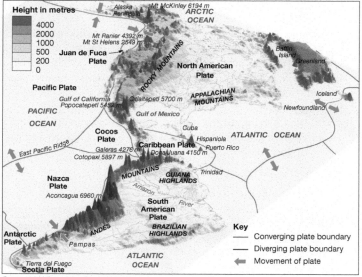

FIGURE 1 Landforms of North, Central and South America (not to scale)

Source: MAPgraphics Pty Ltd, Brisbane

The rifting of Iceland

The Mid-Atlantic Ridge passes through Iceland, where the island is splitting in two different areas (see figure 2). This can be seen where Iceland's volcanoes are located, at the point where the North American Plate is drifting to the west and the Eurasian Plate is drifting to the east (see figure 3). New crust is being formed in a rift below the sea, and eventually water from the Atlantic Ocean will fill the widening and deepening gaps between the separated parcels of land.

FIGURE 2 Rifting in Iceland

Source: MAPgraphics Pty Ltd, Brisbane

The Great Rift Valley, Africa

The Great Rift Valley is in Africa (figure 4). It is about 5000 kilometres long, and stretches from Syria in the north to Mozambique in the south. The valley varies in width from 30 kilometres at its narrowest point to 100 kilometres at its widest. In some places it is a few hundred metres deep; in others it can be a few thousand metres deep.

The Great Rift Valley was created through separation that began 35 million years ago, when the African and Arabian plates began pulling apart in the northern region. About 15 million years ago, East Africa began to separate from the rest of Africa along the East African Rift. The volcanic activity in this region has produced many volcanic mountains, such as Mount Kilimanjaro, Mount Kenya and Mount Elgon.

As these rifts continue to grow, new ocean waters will flow into the valleys, separating the landmasses.

5.8.2 Volcano hotspots

Although most volcanoes are formed on plate boundaries, some are located in the middle of plates, a long way from plate boundaries. These volcanoes have formed above a **hotspot** — a single plume of rising mantle. Volcanoes form as the plates slowly move over the hotspot and, over time, a chain of volcanoes can form. Hotspots are found in the ocean and on continents. Examples include the Hawaiian Islands and many of Australia's extinct volcanoes. In Hawaii, the location of the volcanoes gives a clue to the direction and speed of the plate movement.

FIGURE 3 A chain of volcanoes in Iceland

FIGURE 4 The Great Rift Valley, Africa

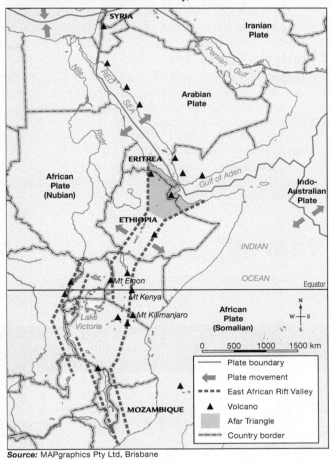

Source: MAPgraphics Pty Ltd, Brisbane

5.8 Activities

To answer questions online and to receive **immediate feedback** and **sample responses** for every question, go to your learnON title at www.jacplus.com.au. *Note*: Question numbers may vary slightly.

Explain

1. Go to the **Hawaii's hotspot** weblink in the Resources tab and explain how hotspot volcanoes form.
2. Refer to figures 2 and 3. Explain why a chain of volcanoes, like the one in the photograph, forms in Iceland. What is happening to the plates?

Discover

3. Use an atlas to find the Cotopaxi volcano. In which country is it located? How high is it?
4. Use an atlas or Google Earth to locate the islands on the Mid-Atlantic Ridge. Give the latitude and longitude for three locations. Describe the *interconnection* between the location of the ridge and the location of islands and volcanoes.

Predict

5. Draw what you imagine Iceland will look like many thousands of years in the future after further rifting. Provide new names for each of the smaller islands. In which direction, and towards which continent, will each island drift? Describe key *changes*.

6. Draw a series of sketches to show what you predict will happen to the African landmass as the Great Rift Valley continues to rift. Include a map of Africa showing the *change* in shape that might occur. You need to annotate your sketches to justify the predictions you have made.

Think

7. Refer to an atlas map of Africa and look at the shape of the island of Madagascar. Try to imagine fitting this island back into the mainland. Using plate tectonic terms, write a paragraph to describe how Madagascar's location has *changed* over time.
8. Describe the *changes* occurring in
   (a) the Great Rift Valley
   (b) Iceland
   that are causing volcanoes to form.
9. How is the *scale* of the *changes* happening in Iceland different from the *scale* of *change* happening in the Great Rift Valley?

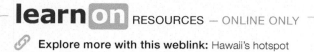
learnon RESOURCES — ONLINE ONLY

🔗 **Explore more with this weblink:** Hawaii's hotspot

 my **World** Atlas **Deepen your understanding of this topic with related case studies and questions.**
   ❯ Hawaii's hot spot

5.9 How did Mount Taranaki form?

5.9.1 Introducing Mount Taranaki, New Zealand

New Zealand's Mount Taranaki is named after the Māori terms *tara* meaning 'mountain peak' and *ngaki* meaning 'shining' (because the mountain is covered with snow in winter).

Mount Taranaki is 2518 metres high and is the largest volcano on New Zealand's mainland. It is located in the south-west of the North Island (see figure 1).

Mount Taranaki was formed 135 000 years ago by subduction of the Pacific Plate below the Australian Plate. It is a stratovolcano — a conical volcano consisting of layers of pumice, lava, ash and tephra. Mount Taranaki is symmetrical, looking the same on both sides of a central point. It is the only active volcano in a chain in this region. The other volcanoes were once very large but have been eroded over time.

The summit of Mount Taranaki is a lava dome in the middle of a crater that is filled with ice and snow. The mountain is considered likely to erupt again. There are significant potential hazards from lahars, avalanches and floods. A circular plain of volcanic material

FIGURE 1 Location of Mount Taranaki on the North Island of New Zealand

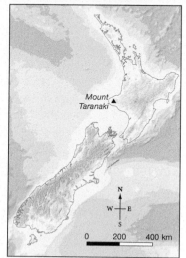

Source: Spatial Vision

surrounding the mountain was formed from lahars (see figure 1 in subtopic 5.11) and landslides. Some of these flows reached the coast in the past. The volcano's lower flanks are covered in forest, and are part of the national park. There is a clear line between the park boundary and surrounding farmland.

FIGURE 2 Mount Taranaki has a near-perfect conical shape

FIGURE 3 Aerial photo of Mount Taranaki

FIGURE 4 Topographic map of Mount Taranaki

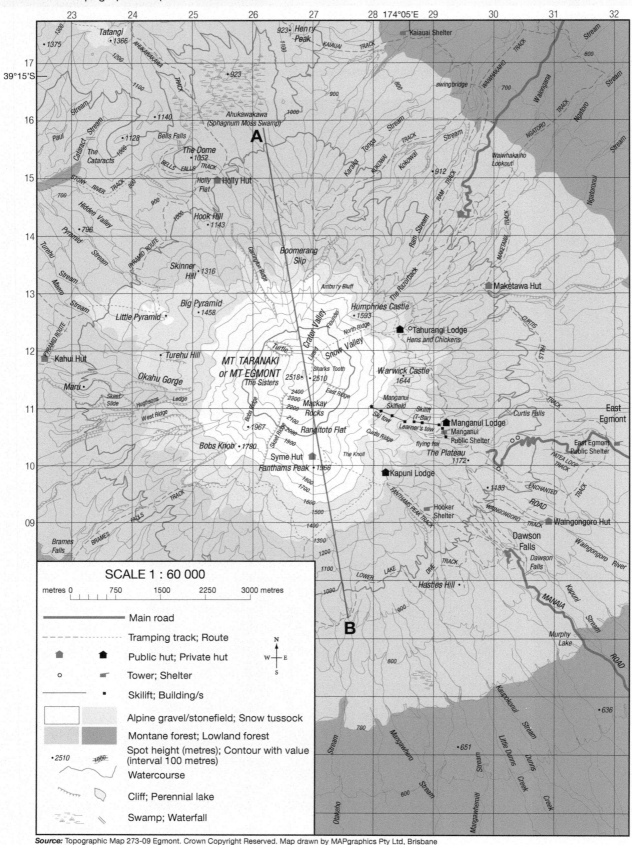

SCALE 1 : 60 000

metres 0   750   1500   2250   3000 metres

Main road

Tramping track; Route

Public hut; Private hut

Tower; Shelter

Skilift; Building/s

Alpine gravel/stonefield; Snow tussock

Montane forest; Lowland forest

Spot height (metres); Contour with value
(interval 100 metres)

•2510   1000

Watercourse

Cliff; Perennial lake

Swamp; Waterfall

N
W—E
S

Source: Topographic Map 273-09 Egmont. Crown Copyright Reserved. Map drawn by MAPgraphics Pty Ltd, Brisbane

5.9 Activities

To answer questions online and to receive **immediate feedback** and **sample responses** for every question, go to your learnON title at www.jacplus.com.au. *Note*: Question numbers may vary slightly.

Remember

1. Where is Mount Taranaki located?
2. What is a stratovolcano?
3. Refer to figure 4.
    (a) What is the grid reference for the spot height of Mount Taranaki?
    (b) Calculate the number of private huts and public huts.
    (c) Name the ski field.
    (d) How many ski tows and lifts are there at the ski field? Calculate the length of each.
    (e) Name and give the grid reference of a lodge in which skiers could stay.
    (f) Name the other two lodges on the map.
    (g) Bushwalking is a popular activity. How many huts are on the map?

Explain

4. Mount Taranaki receives between 3200 mm and 6400 mm of rainfall each year. How would this contribute to the shape of this landform?
5. Describe evidence from the aerial photo in figure 3 that the national park has protected forests around the volcano. (See the 'Interpreting an aerial photo' SkillBuilder in subtopic 5.12.)

Discover

6. Use the **Mt Taranaki Live** weblink in the Resources tab to view Mount Taranaki using the webcam.

Predict

7. Use figures 2, 3 and 4 to describe where you think lava would flow if Mount Taranaki erupted. Describe the potential *changes* to the human and natural *environment*.

learn **on** RESOURCES — ONLINE ONLY

🔗 **Explore more with this weblink:** Mt Taranaki Live

5.10 SkillBuilder: Drawing simple cross-sections

WHAT ARE CROSS-SECTIONS?

A cross-section is a side-on, or cut-away view of the land, as if it had been sliced through by a knife. Cross-sections provide us with an idea of the shape of the land. We can use contour lines on topographic maps to draw a cross-section between any two points.

Go online to access:

- a clear step-by-step explanation to help you master the skill
- a model of what you are aiming for
- a checklist of key aspects of the skill
- a series of questions to help you apply the skill and to check your understanding.

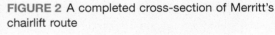

FIGURE 2 A completed cross-section of Merritt's chairlift route

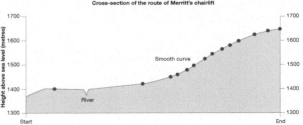

5.11 What is the anatomy of a volcano?

5.11.1 Volcanic eruptions

Volcanic mountains are formed when magma pushes its way to the Earth's surface and then erupts as lava, ash, rocks and volcanic gases. These erupting materials build up around the vent through which they erupted.

A volcanic eruption can be slow or spectacular, and can result in a number of different displays (see figure 1).

5.11.2 Volcanic shapes

Volcanoes come in a variety of shapes and sizes, forming different landforms. There are four main types and each depends on:

- the type of lava that erupts
- the amount and type of ash that erupts
- the combination of lava and ash.

Lava that is rich in silica (a mineral present in sand and quartz) is highly viscous and is thick and slow moving. If the lava is low in silica, it tends to be very runny and may flow for many kilometres before it cools and hardens to become rock. Volcanoes that erupt runny lava tend to have broad, flat sides (shield volcanoes). Those that erupt thick, treacle-like lava tend to have much steeper sides (dome volcanoes).

Heavy ash material, like volcanic bombs, settles close to the crater while lighter ash is carried further away. Volcanoes that are built up through falls of ash are steep-sided cinder cones.

The most common type of volcano is one built up of both ash and lava; this is called a composite volcano.

FIGURE 1 The anatomy of a volcano

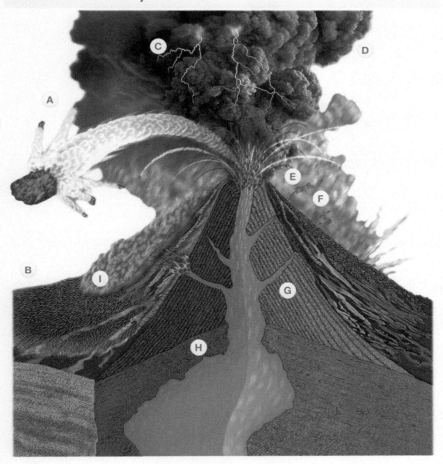

A A fragment of lava greater than 64 millimetres in diameter is called a volcanic bomb. They are often solid pieces of lava from past eruptions that formed part of the cone.

B A pyroclastic flow is a superheated avalanche of rock, ash and lava that rushes down the mountain with devastating effects. The flow can travel at up to 240 kilometres per hour and reach temperatures of 800 °C. When Mount Pelée erupted in 1902, on the island of Martinique in the Caribbean, a pyroclastic flow covered the town of Saint-Pierre, killing all but two of the town's 30 000 inhabitants.

C Lightning is often generated by the friction of swirling ash particles.

D As rock is pulverised by the force of the eruption, it becomes very fine ash, and is carried by wind away from the crater as an ash cloud. Volcanic ash may blanket the ground to a depth of many metres. In the eruption of Mount Vesuvius in 79CE, volcanic ash completely covered two large towns: Pompeii and Herculaneum.

E A volcanic cone is made up of layers of ash and lava from previous eruptions. If the volcano has not erupted for thousands of years (i.e. is dormant), these layers will be eroded away.

F Lava may be either runny or viscous, and can flow for many kilometres before it solidifies, thereby building up the Earth's surface.

G Pressure may force magma through a branch pipe or side vent. In the eruption of Mount St Helens in 1980, in the United States, the side of the mountain collapsed and the side vent became the main vent.

H Where two plates move apart, molten rock from the mantle flows upward into a magma chamber. More rock is melted and erupts violently upwards. Magma is generally within the temperature range of 700 °C to 1300 °C.

I When pyroclastic flows melt snow and ice, and mix with rocks and stones, a very wet mixture called a lahar can form. Lahars can flow quickly down the sides of volcanoes and cause much damage. One lahar that formed in 1985 on Nevado del Ruiz volcano in Colombia travelled at up to 50 kilometres per hour and was up to 40 metres high in some places. A wall of mud, water and debris travelled 73 kilometres to the town of Armero, devastating it. More than 23 000 people died that night and 5000 homes were destroyed.

FIGURE 2 Four volcanic landforms

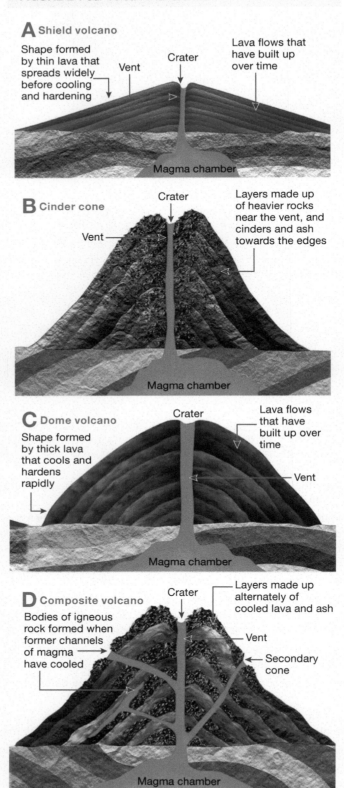

A Shield volcano

Shape formed by thin lava that spreads widely before cooling and hardening

Vent

Crater

Lava flows that have built up over time

Magma chamber

B Cinder cone

Crater

Vent

Layers made up of heavier rocks near the vent, and cinders and ash towards the edges

Magma chamber

C Dome volcano

Shape formed by thick lava that cools and hardens rapidly

Crater

Lava flows that have built up over time

Vent

Magma chamber

D Composite volcano

Bodies of igneous rock formed when former channels of magma have cooled

Crater

Layers made up alternately of cooled lava and ash

Vent

Secondary cone

Magma chamber

5.11 Activities

To answer questions online and to receive **immediate feedback** and **sample responses** for every question, go to your learnON title at www.jacplus.com.au. *Note*: Question numbers may vary slightly.

Remember

1. What are the three main factors that determine the shape of a volcano?

Explain

2. Explain how the different shapes of volcano shown in figure 2 are the results of different materials being ejected.
3. (a) Refer to figure 1. Describe, in detail, the *changes* to the *environment* that volcanic eruptions can cause.
   (b) Which *changes* would impact on a small *scale* and which would impact on a larger *scale*?

Think

3. Use the internet to find pictures of volcanic landforms and materials. These include crater lakes, geysers, calderas, fields of ash deposits, volcanic plugs, lava tubes, hummocks and pumice. You could also find pictures of the two types of lava: a'a and pahoehoe. Use your pictures to put together a field guide to volcanic landforms. Each page should contain a picture of the landform, a brief description and a place where it could be found — sometimes they are tourist attractions.

---

 myWorldAtlas — Deepen your understanding of this topic with related case studies and questions.
◉ Lahars

5.12 SkillBuilder: Interpreting an aerial photo

WHAT ARE AERIAL PHOTOS?

Aerial photographs are those that are taken from above the Earth from an aircraft. **Oblique aerial photos** are those taken from an angle from an aircraft. **Vertical aerial photos** are taken from directly above; that is, looking straight down onto objects. Aerial photos can reveal details that are not recorded on maps. It is easy to see landforms with distinct shapes, different landscapes, land uses, specific places and spatial patterns of the environment.

Go online to access:

- a clear step-by-step explanation to help you master the skill
- a model of what you are aiming for
- a checklist of key aspects of the skill
- a series of questions to help you apply the skill and to check your understanding.

FIGURE 1 Cartographers use different types of photographs

Satellite imagery

Vertical aerial photographs

Oblique aerial photographs

Ground-level photographs

5.13 How do volcanic eruptions affect people?

Access this subtopic at **www.jacplus.com.au**

 **Deepen your understanding of this topic with related case studies and questions.**
 ❯ Mt Vesuvius

5.14 Review

5.14.1 Review

The Review section contains a range of different questions and activities to help you revise and recall what you have learned, especially prior to a topic test.

5.14.2 Reflect

The Reflect section provides you with an opportunity to apply and extend your learning.

Access this subtopic at **www.jacplus.com.au**

TOPIC 6
Rainforest landscapes

6.1 Overview

Numerous **videos** and **interactivities** are embedded just where you need them, at the point of learning, in your learnON title at www.jacplus.com.au. They will help you to learn the content and concepts covered in this topic.

6.1.1 Introduction

What do you know about rainforest landscapes? Did you know that rainforests have the greatest bio-diversity of any forest environment? They contain complex layers that support thousands of species of plants and animals. The rainforest has supplied resources to all people, including indigenous communi-ties. People are concerned that clearing large areas of this landscape is creating negative impacts that are unsustainable.

Aerial view of the Amazon rainforest in Peru, South America

1. Create a mind map of all the words that come to your mind when you hear the word *rainforest*. Place these words into specific categories such as *animals, plants, colours* or *structure*.
2. Using magazines, newspapers, the internet and other resources, create a collage that represents a rainforest landscape to you. What emotions does this image make you feel?
3. Refer to the image on these pages. Why do we call rainforests a 'green landscape'?
4. Try to list the things in your home that might originally have come from a rainforest. Have they *changed* from their original form? How do you think they got here?
5. Think of your favourite *place*, the place you most like visiting. How would you feel if that place was *changed* or destroyed and you could not visit or use it in ways you used to? What could you do about this?

6.2 What are the characteristics of a rainforest?

6.2.1 Rainforests

Forests that grow in constantly wet conditions are defined as rainforest landscapes. A rainforest is an example of a biome (a community of plants and animals spread over a large natural area). Rainforests are located wherever the annual rainfall is more than 1300 millimetres and is evenly spread throughout the year. While tropical rainforests are the best known of these landscapes, there are also other types.

6.2.2 Tropical rainforests and their processes

Tropical rainforest landscapes are found where there are both high temperatures and high precipitation. The sun's rays that reach the Earth near the equator have a smaller area of the Earth and atmosphere to heat than rays reaching the Earth at higher latitudes. Therefore, it is hotter at the equator than at higher latitudes. Rainforests are also generally warmer at night, because the cloud cover and high humidity help to keep the heat in. Tropical rainforests have a hot climate right throughout the year with no summer or winter. High precipitation around the equator is mainly due to convectional rainfall and is often associated with thunderstorms. Convectional rainfall occurs when warm, moist air is heated when it moves over a hot surface on Earth. As the air is heated it expands and becomes lighter than the surrounding air. This causes it to rise. If the air continues to rise, condensation and precipitation occur. This combination of high temperatures and high precipitation influences the global distribution of the tropical rainforest landscape. Plants flourish in these rainforests, which support a huge number of plants and animals — perhaps as many as 90 per cent of all known **species**. Poison-dart frogs, birds of paradise, piranha, tarantulas, anacondas, Komodo dragons and vampire bats are all found in tropical rainforests.

Tropical rainforests that occur in the mountains, 1000 metres or more above sea level, are called montane rainforests. Other tropical rainforests are known as lowland rainforests (see figure 1).

Lowland tropical rainforest

Lowland tropical rainforests form the majority of the world's tropical rainforests. They grow at elevations generally below 1000 metres. Trees in lowland forests are usually taller than those in montane forest and include a greater diversity of fruiting trees. These attract animals and birds adapted to feed on their fruits. These rainforests are far more threatened than montane forests because of their accessibility, soils that are more suitable for agriculture and more valuable hardwoods for timber. Lowland forests occur in a belt around the equator, with the largest areas in the Amazon Basin of South America, the Congo Basin of central Africa, Indonesia and New Guinea.

FIGURE 1 World rainforest types

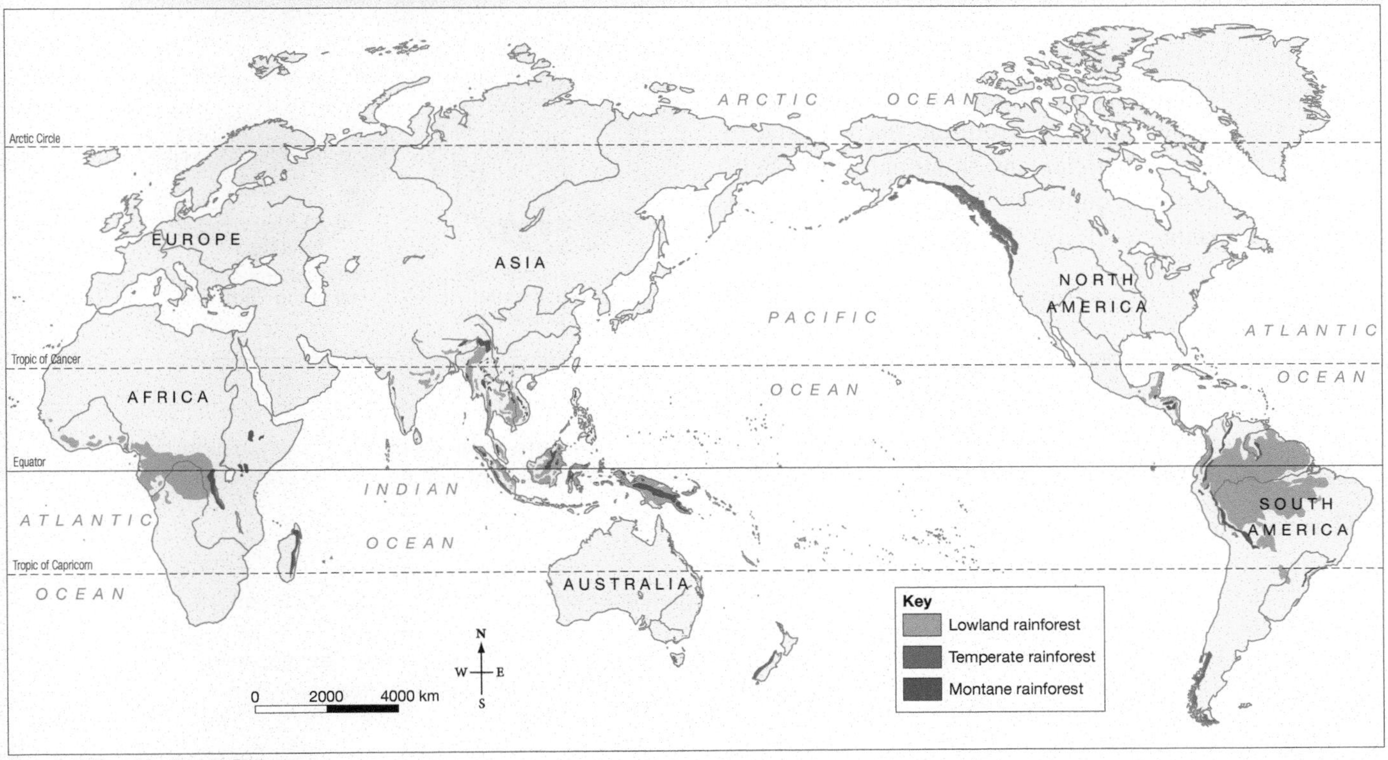

Key
- Lowland rainforest
- Temperate rainforest
- Montane rainforest

Source: MAPgraphics Pty Ltd, Brisbane

FIGURE 2 (a) Montane rainforest, (b) temperate rainforest and (c) lowland rainforest

Temperate rainforests

The large area of the globe between the tropics and the polar regions (areas within the Arctic and Antarctic circles) is called the **temperate** zone, and rainforests can grow there too. Temperate rainforests occur in North America, Tasmania, New Zealand and China. Giant pandas, Tasmanian devils, brown bears, cougars and wolves all call temperate rainforests home.

6.2.3 Physical processes of a rainforest

Rainforest landscapes are the result of the interaction between the Earth's four main systems or spheres. For example, the trees in a tropical rainforest (biosphere) rely on high levels of precipitation (hydrosphere), warm temperatures (atmosphere) and stability provided by soil (lithosphere) to thrive. Energy from the sun is stored by plants (biosphere). When humans or animals (biosphere) eat the plants, they acquire the energy originally captured by the plants.

FIGURE 3 The Earth's four main systems

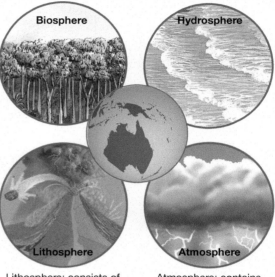

Biosphere: the collection of all Earth's life forms

Hydrosphere: 97 per cent of the Earth's water is found in salty oceans, and the remainder as vapour in the atmosphere and as liquid in groundwater, lakes, rivers, glaciers and snowfields.

Lithosphere: consists of the core, mantle and crust of the Earth

Atmosphere: contains all of the Earth's air

6.2 Activities

To answer questions online and to receive **immediate feedback** and **sample responses** for every question, go to your learnON title at www.jacplus.com.au. *Note:* Question numbers may vary slightly.

Remember

1. What conditions do rainforest *environments* thrive in?
2. What are the differences between montane and lowland rainforest *environments*? What causes these *changes* in rainforest type?
3. Describe the distribution of rainforests around the world. Think about in which continents and between which latitudes they are found, the size and *scale* of them, and whether they are continuous or scattered.

Explain

4. Why are lowland rainforest *environments* more threatened by human activity than montane rainforests?
5. Why are montane forests often called 'cloud forests'?

Discover

6. Refer to figure 1.
   (a) Use an atlas to help you name six countries in the Asia–Pacific region that contain rainforests.
   (b) What type of rainforest *environment* is found:
      (i) in north-eastern Australia
      (ii) along the western coastline of Canada?

Think

7. Refer to figure 3. List Earth's four spheres. Give several examples of features in each sphere.
8. List some Earth sphere interactions from your own daily activities.
9. Why are rainforest *environments* able to support a large range of animals and plants?

6.3 What is a rainforest ecosystem like?

6.3.1 Rainforest ecosystems

Rainforests are unique **ecosystems** consisting of four different layers — the emergent, canopy and understorey layers and the forest floor. Each layer can be identified by its distinct characteristics. Rainforests are actually a community of plants and animals working together to survive, linked in a food web (see figure 2).

Emergents

These are the tallest trees, ranging in height from 30 to 50 metres. They are so named because they rise up or emerge out of the forest canopy. Huge crowns of leaves and abundant animal life thrive on plenty of available sunlight.

Canopy

This describes the array of treetops that form a barrier between the sunlight and the underlying layers. Their height can vary from 20 to 45 metres. This layer contains a distinct **microclimate** and supports a variety of plants and animals. The taller trees host special vines called lianas that intertwine the branches. Other plants called epiphytes use the tree trunks and branches as anchors in order to capture water and sunlight.

Understorey

This layer contains a mixture of smaller trees and ferns that receive only about five per cent of the sun's energy. Many animals move around in the darkness and humidity, using the vines as highways.

Forest floor

This bottom layer is dominated by a thick carpet of leaves, fallen trees and huge buttress roots that support the giant trees above. Rainforest soils give the impression of being fertile because they support an enormous number of trees and plants. However, this impression is wrong, as the soil in rainforests is generally poor. Leaves and other matter are recycled by the many organisms to

FIGURE 1 Layers in a tropical rainforest

Canopy

Tall emergent tree

Liana

Buttress roots

Epiphytes

Ferns

Undergrowth

Moss

create a living **compost**. The roots of trees must 'snatch' these nutrients from the soil before heavy rains wash them away and they are lost through a process called **leaching**.

Larger animals also roam through this layer in search of food.

FIGURE 2 An example of a typical food web in an Australian rainforest

6.3 Activities

To answer questions online and to receive **immediate feedback** and **sample responses** for every question, go to your learnON title at www.jacplus.com.au. *Note:* Question numbers may vary slightly.

Remember

1. How many layers are there in a rainforest *environment*?
2. What are the tallest trees in the rainforest called?

Explain

3. Describe how conditions in the canopy layer differ from those on the forest floor.
4. Draw up and complete a table like the one below that summarises the features of a rainforest *environment*.

Layer	Height	Amount of light	Features

Discover

5. Many rainforest animals live their whole life in the trees. Using the internet to help you, give some examples of these animals and conduct research into the habits of one animal.
6. Use the **Treehouse** weblink in the Resources tab (click on the picture, then select the *Jewels of the Earth* activity) to explore the layers of the rainforest and the plants and animals that inhabit them.

Think

7. In a rainforest, the soil below the trees is often poor and shallow, and the trees create their own nutrients. In one sentence describe how this happens, and draw a labelled sketch to illustrate the process.
8. What *change* might you expect in the success of plant growth if the rainforest trees are removed and crops are planted instead? Why?
9. Imagine you are a raindrop. Recreate your journey through a rainforest, passing through each of the forest layers. Read or act out your descriptions to the rest of the class.
10. Identify key characteristics of a tropical rainforest.

 **learnon** RESOURCES — ONLINE ONLY

 Explore more with this weblink: Treehouse

6.4 SkillBuilder: Creating and describing complex overlay maps

WHAT IS A COMPLEX OVERLAY MAP?

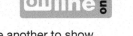

A complex overlay map is created when one or more maps of the same area are laid over one another to show similarities and differences between the mapped information. Traditionally, the second map is on tracing paper that is attached to the original page.

Complex overlay maps show relationships between factors — the similarities and the differences in a pattern.

Go online to access:

- a clear step-by-step explanation to help you master the skill
- a model of what you are aiming for
- a checklist of key aspects of the skill
- a series of questions to help you apply the skill and to check your understanding.

FIGURE 1 An illustration of a completed complex overlay map showing Australia's seasonal rainfall patterns (left), drainage catchments (centre) and average annual rainfall (right)

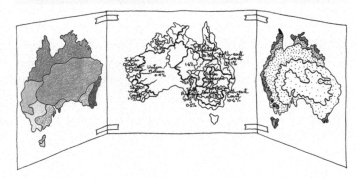

learnon RESOURCES — ONLINE ONLY

Watch this eLesson: Creating and describing complex overlay maps (eles-1656)

Try out this interactivity: Creating and describing complex overlay maps (int-3152)

6.5 How have Australia's rainforests changed?

6.5.1 Australian rainforests

Although hard to believe now, Australia was once mostly covered in rainforest! Even areas that today are deserts were once teeming with plant and animal life similar to those in the Amazon. This is because Australia was further north than it is today. Over the past 100 million years, however, a series of events has gradually reduced the area of Australia's rainforests (see figure 1).

FIGURE 1 The development of Australian rainforests through time

Triassic period (245–208 mya)
- Foundation of Australian rainforests was occurring as massive volcanic eruptions laid down thick layers of ash near freshwater lakes.

Early Cretaceous period (144–95 mya)
- Australia, as part of Gondwanaland, was very close to the South Pole and experienced the same conditions as present polar regions. However, the climate was temperate, supporting plant life, and the first flowering plants appeared.

Late Cretaceous period (95–66.4 mya)
- The climate was temperate and humid; species found in rainforests today began to develop. Dinosaurs were still present.

Present

| 260 | 240 | 220 | 200 | 180 | 160 | 140 | 120 | 100 | 80 | 60 | 40 | 20 |

[Dates are expressed in mya — million years ago.]

Jurassic period (208–144 mya)
- Ash deposits developed into sedimentary rock; plants such as the cycad and southern pine first appeared.

Late Oligocene/early Miocene periods (25–20 mya)
- The whole of Australia was covered by rainforest at the start of this period, but by the end of the Miocene sclerophyll forest and grasslands were starting to emerge. The rainforests would have been similar to those of today.

The gradual movement of Australia southwards as it separated from Gondwanaland and a series of **ice ages** have combined to make it a drier place (see figure 2). Rainforests have become confined mainly to the mountains and **gorges** of the Great Dividing Range and Tasmania. These areas have higher rainfall and fewer fires.

The farming practices of European immigrants have reduced much of the remaining rainforest — in the past 200 years, more than 70 per cent of these forests have been cleared.

Scientists have identified three major types of rainforest in Australia (see figure 3). There are examples of all three in Queensland. This diversity occurs nowhere else on Earth.

Much of Australia's tropical rainforests are now World Heritage areas. This means they have been listed by UNESCO as being of global importance. The Wet Tropics of Queensland are a World Heritage area containing some of the oldest rainforests in the world. They have the world's highest concentration of flowering plants, and have records that show Aboriginal communities are the world's oldest indigenous rainforest culture.

The indigenous inhabitants of the Daintree Rainforest in North Queensland are the Kuku Yalanji Aboriginal people, believed to have lived in this area for more than 9000 years by European estimates. Their culture is uniquely adapted to the rainforest environment.

FIGURE 2 Difference in the location of Australia when it formed part of Gondwanaland compared with its location today

Gondwanaland   **Present day**

 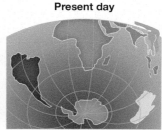

FIGURE 3 Australia's rainforest areas

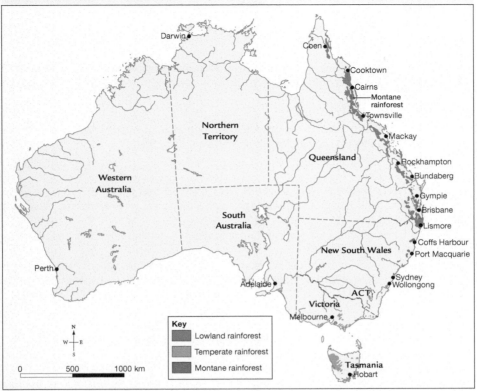

Source: MAPgraphics Pty Ltd, Brisbane

For the Kuku Yalanji, the natural world is often thought of in human terms and is closely linked to the people. Any changes to the environment are seen as changes to themselves. Because of the powerful properties attributed to most story places (sites with links to the Dreamings) of the Daintree, the Kuku Yalanji regard damage and destruction to the environment as unacceptable.

The Kuku Yalanji people gather their food and medicine and many of their implements, weapons, fibres and construction material

FIGURE 4 Rainforest plants were used to make goods such as these baskets that were used for storage, food collection, carrying personal possessions, and leaching poisons (from seeds) in fresh running water.

from plants in their environment. The natural patterns and cycles of the rainforest give important information about the food that is available. The plants are their calendar, marking the seasons. For example, when blue ginger (jun

jun) is fruiting it is time to catch scrub turkey (diwan), and when mat grass (jilngan) is flowering it is time to collect the eggs of the scrub fowl (jarruka).

6.5 Activities

To answer questions online and to receive **immediate feedback** and **sample responses** for every question, go to your learnON title at www.jacplus.com.au. *Note:* Question numbers may vary slightly.

Remember

1. List the reasons given in this subtopic for the gradual disappearance of Australia's rainforest *environments*.
2. What resources does the rainforest provide for the Kuku Yalanji people?

Explain

3. How has the *scale* of Australian rainforest *environments changed* over time?
4. Describe the location and distribution of Australia's remaining rainforest *environments*. What factors have contributed to their survival here?
5. There are three major types of rainforest *environments* found in Australia. What makes Queensland's rainforests unique? Why is this possible?
6. Refer to figure 3. Why are there no rainforest *environments* on the western side of Australia?
7. Why do the Kuku Yalanji people regard damage to the Daintree Rainforest as unacceptable?

Discover

8. Use the **UNESCO Heritage** weblink in the Resources tab to complete the following.
   (a) On a map of Australia, locate and label Australia's World Heritage sites.
   (b) Which three sites have been added most recently?
   (c) Which two sites protect Australian rainforests?
   (d) The Wet Tropics of Queensland are particularly special because they border another World Heritage site. What is this other site?
   (e) What criteria does UNESCO use to determine whether a natural region should be placed on its list?

Think

9. Based on the history of Australia's rainforests and the protection now in place for the remaining forests, what do you think the future holds for this important resource?
10. A hotel chain has applied to the Queensland government for permission to build a resort in the Daintree. Assess this proposal from the perspectives of the developers, government, local residents, environmentalists and Kuku Yalanji people. Try to make a decision as to whether this project should be approved. This could be completed in small groups or debated as a class.

 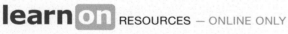 RESOURCES — ONLINE ONLY

🔗 **Explore more with this weblink:** UNESCO Heritage

6.6 How is the Amazon rainforest changing?

6.6.1 The Amazon rainforest

The world's largest remaining rainforest is in the Amazon **Basin** in South America. This truly remarkable forest is under increasing threat from forestry, mining and farming. The loss may cause severe problems worldwide. Most of us use rainforest products every day. More importantly, however, rainforests help control the world's climate and our oxygen supply. So the next time you eat chocolate, treat your asthma, play a guitar or even take a deep breath, you should thank the Amazon rainforest.

FIGURE 1 The Amazon Basin

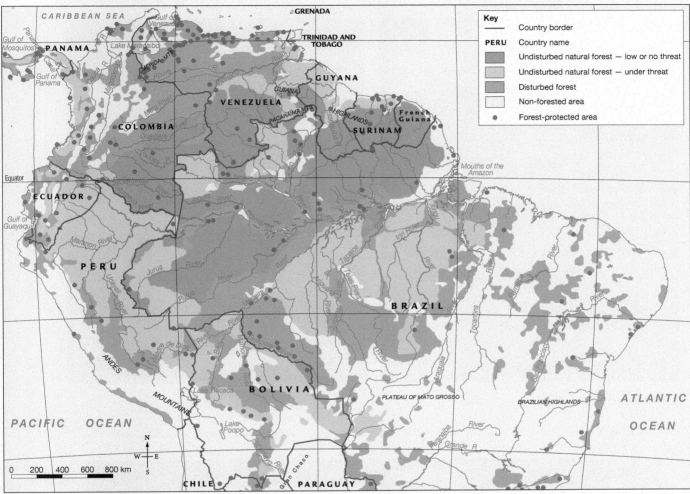

Key

——	Country border
PERU	Country name
▨	Undisturbed natural forest — low or no threat
▨	Undisturbed natural forest — under threat
▨	Disturbed forest
☐	Non-forested area
•	Forest-protected area

Source: MAPgraphics Pty Ltd, Brisbane

Ⓐ The Amazon River and those rivers that feed into it (tributaries) contain one-fifth of the world's fresh water, and more than 2000 species of fish — more than in the Atlantic and Pacific oceans combined.

Ⓑ The mouth of the Amazon River is approximately 325 kilometres wide and contains an island the size of Switzerland!
- The Amazon forest is home to more than 40 000 species of plants, 1300 bird species, 430 different mammals and 2.5 million different insects.
- Approximately 1.3 million tons of sediment is transported by the Amazon River to the sea daily.
- No bridges cross the main trunk of the Amazon River, which locals call the Ocean River.
- Since 2000, the Amazon rainforest has been facing deforestation at an average rate of 50 football fields per minute.
- The Amazon is the second longest river in the world, but it carries more water than the next six largest rivers combined.
- The Amazon River drains nearly 40 per cent of South America.
- There are official plans for 412 dams to be in operation in the Amazon River and its headwaters.
- Since 1900, more than 90 indigenous groups have disappeared in Brazil alone.

FIGURE 2 The brown waters of the Amazon show that it is carrying a lot of sediment.

FIGURE 3 Area cleared for ranching in the Amazon rainforest

FIGURE 4 Development is clearly visible within the green carpet of the Amazon rainforest between 1975 (left) and 2012 (right).

Source: NASA/Landsat

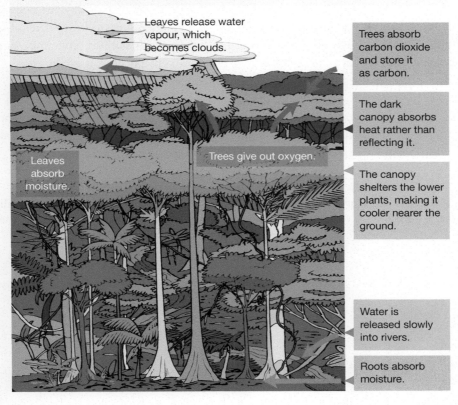

FIGURE 5 Rainforests play a vital role in controlling the world's climate and oxygen supply. Scientists believe that half of all the world's oxygen is produced by the Amazon rainforest alone.

Leaves release water vapour, which becomes clouds.

Trees absorb carbon dioxide and store it as carbon.

The dark canopy absorbs heat rather than reflecting it.

Leaves absorb moisture.

Trees give out oxygen.

The canopy shelters the lower plants, making it cooler nearer the ground.

Water is released slowly into rivers.

Roots absorb moisture.

6.6.2 Amazing rainforests

- More than 7000 modern medicines are made from rainforest plants. They can be used to treat problems from headaches to killer diseases like malaria. They are used by people who suffer from multiple sclerosis, Parkinson's disease, leukaemia, asthma, acne, arthritis, diabetes, dysentery and heart disease among many others.
- Even animals can be used to cure human diseases. Tree frogs from Australia give off a chemical that can heal sores, and a similar chemical from a South American frog is used as a powerful painkiller.
- The poisonous venom from an Amazonian snake is used to treat high blood pressure.
- Only one per cent of the known plants and animals of the rainforest have been properly analysed for their medicinal potential. Perhaps the greatest benefits to medicine and our own health, however, are yet to come.
- Rainforests are home to the greatest profusion of life on the planet: at least half of all known plants and animals live in rainforests.
- At least 50 million indigenous peoples live in rainforests worldwide. From the Kuna people of Panama and the Yanomami of Brazil to the Baka people of Cameroon and the Penan of Borneo (Indonesia), these people have traditionally lived a way of life that has little impact on their forest home.

FIGURE 6 Skin secretions from frogs such as the Waxy Monkey Treefrog (*Phyllomedusa bicolor*) contain powerful painkillers.

- The people who live in or near the rainforests gain much of their food from the forest. But rainforests also supply the supermarkets of the world with their bounty. Most of these fruits and nuts are now grown by farmers rather than harvested directly from the forest, but it was in the rainforests that they originated.
- Chocolate first came from cacao trees native to the Amazon rainforest. Today the cocoa in the chocolate you eat is most likely to have come from huge cacao plantations in West Africa. Similarly, brazil and cashew nuts, cinnamon, ginger, pepper, vanilla, bananas, pineapples, coconuts, paw-paws, mangoes and avocados were all originally rainforest plants. Even the gum used in chewing gum comes from a rainforest plant, as does the tree that produces rubber.
- Rainforest trees are generally hardwood trees, making them resistant to decay and attractive for building. Well-known rainforest timbers are mahogany, teak, ebony, balsa and rosewood. Rosewood is particularly interesting, as it is considered the best timber in the world for guitar making. In many tropical countries, people also collect timber as fuel for cooking or heating.

FIGURE 7 The Kamayurá people of the Brazilian rainforest live a traditional way of life.

FIGURE 8 Food products such as chocolate and chewing gum are made from ingredients that originally came from the rainforest.

6.6 Activities

To answer questions online and to receive **immediate feedback** and **sample responses** for every question, go to your learnON title at www.jacplus.com.au. *Note:* Question numbers may vary slightly.

Explain

1. Which of the present uses of the rainforest do you think is the most *sustainable* for the forest's future? Explain your answer.
2. Refer to figure 4. Why does the clearing and *change* in the Amazon appear to occur in straight lines?
3. Looking at figures 2, 3 and 4 for *interconnections*, what do you think could be contributing to the high levels of sediment in the Amazon River? Why?
4. Refer to figure 5.
   (a) Explain the role of the rainforest *environment* in relation to the climate.
   (b) Why are rainforests sometimes called 'the lungs of the Earth'?

Discover

5. Look carefully at figure 1.
   (a) List the countries of South America into which the Amazon rainforest extends.
   (b) Which country contains most of the Amazon rainforest?
   (c) Why do you think there are so few large cities in the rainforest?
   (d) Estimate the percentage of the rainforest that can be considered:
      (i) under low or no threat
      (ii) under threat
      (iii) disturbed.
   Describe in your own words what each of these terms means.

6. Using a piece of tracing paper, trace the Amazon River and its tributaries. Draw a single line that joins the source of each of the tributaries. Shade the area within this line using a light blue pencil: this area is known as the **catchment**, or basin, of the river. Overlay your completed diagram on the map of the forest and comment on the *interconnection* between the river and the forest.

7. This subtopic lists only a few of the products we use from rainforests. List the value of these and other rainforest products under the following headings.
   (a) Valued by different cultures
   (b) Valued economically
   (c) Valued for its aesthetic value (beauty)
   (d) Other

8. Use the **Treehouse** weblink in the Resources tab (click on the picture then choose the *Track it back* activity) to learn how the food you eat comes from the rainforest.

9. Use the **Amazon tour** weblink in the Resources tab to take a tour through an Amazon rainforest slideshow.

Think

10. If development in the Amazon Basin continues as seen in figures 3 and 4, what could be the consequences in terms of the processes shown in figure 5?

11. Make a list of things in your home that may come from the rainforest *environment*. Remember to look in the medicine cupboard and the pantry as well as at the furniture. Perhaps you could bring some examples to school and your class could set up a display.

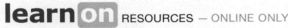 RESOURCES — ONLINE ONLY

 Explore more with these weblinks: Treehouse, Amazon tour

6.7 SkillBuilder: Drawing a précis map

WHAT IS A PRÉCIS MAP?

A précis map is a simplified map — the cartographer has decided which details to leave in and which to leave out. It is different from a sketch map, which includes all the main features.

Go online to access:
- a clear step-by-step explanation to help you master the skill
- a model of what you are aiming for
- a checklist of key aspects of the skill
- a series of questions to help you apply the skill and to check your understanding.

FIGURE 1 Précis map showing areas where the Penan live in Sarawak, in Malaysian Borneo

Source: MAPgraphics Pty Ltd, Brisbane

6.8 How do indigenous peoples use rainforests?

Access this subtopic at **www.jacplus.com.au**

6.9 Why are rainforests disappearing?

6.9.1 Factors causing rainforest deforestation

Rainforests have the potential to provide a wide variety of useful resources. The temptation to use these pristine areas is often too difficult for people to resist, especially if they live in poverty. As a result, all around the world, rainforests are being destroyed for economic gain. The main reasons for rainforests being cleared are described below.

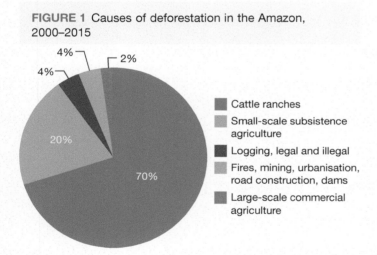

FIGURE 1 Causes of deforestation in the Amazon, 2000–2015

- 4%
- 4%
- 2%
- 20%
- 70%

- ■ Cattle ranches
- ■ Small-scale subsistence agriculture
- ■ Logging, legal and illegal
- ■ Fires, mining, urbanisation, road construction, dams
- ■ Large-scale commercial agriculture

FIGURE 2 It is thought that up to 70 per cent of logging in Brazil and Indonesia could be illegal.

Commercial logging

There are two main types of logging: **clearfelling** and **selective logging**. When a forest is clearfelled, all trees are removed either by chainsaw or with heavy machinery such as bulldozers. In selective logging, only the best and most valuable trees are cut down. But in clearing forest to reach those trees, it is estimated that a hectare (10 000 square metres) of forest is destroyed for each log removed.

Farming

Rainforests grow in many developing countries. These countries struggle to provide the basic necessities of life for their people, and their populations are often rapidly increasing in size. In these countries, the land on which the forest grows is seen as more valuable than the forest itself.

Highways create access to these areas, opening up parts of the rainforest once almost impossible to reach. Soon after the roads are built, settlers (called homesteaders) arrive. Claiming a piece of the forest that borders the road, the homesteaders chop down a few trees as timber for fencing or a house, and then sets fire to the rest.

Once the initial 'land rush' is over and all the land beside the roads has been claimed, tracks and roads leading from the highways will push deeper and deeper into the forest. Soon an area of 50 kilometres either side of the highway will have been destroyed and replaced by small farms or large-scale commercial farms that raise beef or crops for export to the richer countries of the world.

FIGURE 3 Blocks of rainforest in Peru are burned to clear the area for agricultural use — here, maize seedlings have been planted in the clearing.

Mining

Many rainforests are growing on land that also contains large energy and mineral deposits such as oil, gold, silver, bauxite, iron ore, copper and zinc. Mineral companies build roads to the deposits and set up large-scale mining and processing plants. These plants require large amounts of electricity, and this is often supplied by burning trees to create charcoal or by constructing vast **hydroelectric dams**.

Deep in the Brazilian rainforest, a 2000-square-kilometre dam has been constructed to provide electricity for aluminium smelters. The dam flooded the entire tribal lands of two native peoples, and is so large that it has altered the climate in the area, making it drier.

Another problem created by mining is the pollution of nearby rivers and streams from chemicals used in the processing plants. Rivers downstream from a vast goldmine in Papua New Guinea have been found to contain four times the safe limit of cyanide in the water. Cyanide is used to extract gold from rock.

FIGURE 4 The Ok Tedi gold and copper mine in the Papua New Guinea rainforest. The damage that mining has caused to the surrounding environment can be clearly seen.

6.9 Activities

To answer questions online and to receive **immediate feedback** and **sample responses** for every question, go to your learnON title at www.jacplus.com.au. *Note:* Question numbers may vary slightly.

Remember

1. What is the difference between clearfelling and selective logging?
2. List four *changes*/problems caused by mining operations in rainforests.

Explain

3. 'Many homesteaders are unable to make a good living from the poor tropical soils.' Explain the reasoning behind this statement. You may like to revisit subtopic 6.3 to help you with your response.

Discover

4. Refer to figure 1. What percentage of deforestation is caused by agriculture in the Amazon?
5. Using the internet, research any economic activities that are supported by Australian rainforests.

Predict

6. As a class, discuss the potential long-term problems that could result from the continued commercial use of rainforest *environments* around the world. Develop a list of the top five potential problems.

Think

7. Many rainforest *environments* are located in developing countries. Why does this make the problem of rainforest destruction harder to solve?
8. Mining companies insist that mining in poorer countries brings benefits to the local community. Outline some of the benefits that mining could bring to poorer communities. Then outline any problems that mining could cause for the such communities. What conclusions can you reach?

6.10 How does deforestation affect the environment and people?

6.10.1 Impacts of rainforest deforestation

Deforestation of rainforests around the world is the major cause of problems in this ecosystem. The loss of unique **habitats** is the primary reason species are becoming endangered. Clearing creates smaller islands of vegetation, making it more difficult for animals to communicate and breed. People are also affected by the removal of the rainforest. While indigenous peoples may feel the effects first, others also experience negative consequences.

- About one hectare of rainforest is destroyed every second: this is about twice the size of a soccer pitch.
- Scientists estimate that 137 plants and animals are made extinct daily: that's 50 000 each year. Some haven't even been discovered yet!
- It is believed that in the year 1500 up to nine million indigenous peoples lived in the Amazon rainforest. The number is now lower than 200 000.
- The world loses about two per cent of its rainforest each year, but rates differ between countries.

6.10.2 Impacts on plants and animals

Islands in the forest

Many forests are cleared using fire. These fires will release millions of tonnes of carbon dioxide into the air, increasing the threat of global warming. At the same time, destroying the trees robs the planet of the natural system that helps regulate the amount of carbon dioxide in the air.

In many areas where forests are cleared, it has become a practice to leave behind 'islands' of rainforest. This is meant to assist in the natural regeneration of the forest and also to leave sufficient areas of the natural habitats of plants and animals that live in the rainforest. But is this working?

The islands that are left are often not big enough to ensure the survival of the large numbers of species that live there. For example, the endangered Queen Alexandra's Birdwing (the world's largest butterfly) is facing extinction as its distribution is being condensed into seven isolated blocks of rainforest measuring approximately 1–2 square kilometres in northern Papua New Guinea. These remaining refuges are threatened by surrounding palm oil plantations.

And there are other problems. When the forest is cleared, the exposed earth can quickly erode as the tree roots no longer hold the soil together, making the regrowth of vegetation slow. On steep slopes this can increase the risk of landslides, and sediments can flow into rivers.

During drought, the bare ground can become hot and barren. With the removal of the forest cover there is little moisture stored in the ground and a much lower rate of **evapotranspiration**. This in turn affects the water cycle, reducing the amount of rain that falls on the remaining islands of rainforest, and they quickly dry out.

FIGURE 1 The wingspan of the Queen Alexandra's Birdwing can reach 30 centimetres.

FIGURE 2 Leftover pockets of rainforest are at risk from reduced rainfall and cannot survive drought conditions.

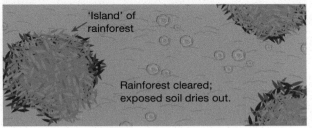

'Island' of rainforest

Rainforest cleared; exposed soil dries out.

(a) Rainforest trees are cleared, with 'islands' left for regeneration.

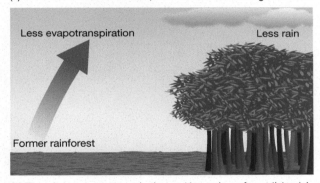

Less evapotranspiration

Less rain

Former rainforest

(b) There is less evapotranspiration and less rain on forest 'islands'.

6.10.3 Indonesia

CASE STUDY

Deforestation in Indonesia and the orangutan

Nearly 10 per cent of the world's rainforests and 40 per cent of all Asian rainforests are found in Indonesia. Less than half of Indonesia's original rainforest area remains. Much of this is in Kalimantan, on the island of Borneo. Forests have been cleared for timber, for plantation crops such as palm oil trees, and to make way for Indonesia's growing population, which is now more than 200 million. Fires lit to clear land in 1982 and 1997 resulted in wildfires that severely damaged large areas of rainforest in Kalimantan. Orangutans, Sumatran tigers and Javan hawk-eagles may disappear from Indonesia as their natural habitats disappear.

Orangutans are the largest tree-living mammals and the only great ape that lives in Asia. They survive only on the islands of Borneo and Sumatra. Current estimates are that orangutans have lost 80 per cent of their habitat in the last 20 years. In 1997–98, wildfires burned through nearly two million hectares of land in Indonesia, killing up to 8000 orangutans.

It is estimated that orangutan numbers have declined by more than 50 per cent in the last 60 years. The current orangutan population is believed to range between 45 000 and 69 000.

FIGURE 3 Mother and baby orangutan

FIGURE 4 Orangutan distribution in Borneo, 1930–2016

Key
Orangutan distribution

1930   1999   2015

BRUNEI   BRUNEI   BRUNEI
MALAYSIA   MALAYSIA   MALAYSIA

B O R N E O   *B O R N E O*   *B O R N E O*   Equator

INDONESIA   INDONESIA   INDONESIA

N
W   E
S

0   250   500 km

Source: IUCN Red list

FIGURE 5 Rainforest distribution in Borneo, 1950–2020

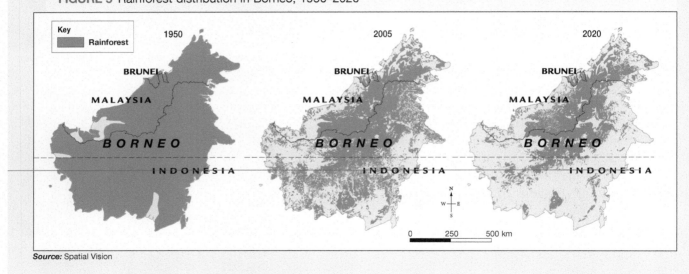

Key
Rainforest

1950   2005   2020

BRUNEI   BRUNEI   BRUNEI
MALAYSIA   MALAYSIA   MALAYSIA

B O R N E O   *B O R N E O*   *B O R N E O*

I N D O N E S I A   I N D O N E S I A   I N D O N E S I A

N
W   E
S

0   250   500 km

Source: Spatial Vision

myWorldAtlas   Deepen your understanding of this topic with related case studies and questions.
  ◉ Orangutans

6.10.4 Impacts on people
Indigenous peoples

As forests are cleared and new occupiers move into the region, the indigenous peoples of the area are often displaced and their cultures may disappear. The homesteaders bring new diseases to which indigenous peoples have no natural immunity. One group, the Nambiquara of Brazil, lost half its population to illness when a road was placed through their tribal land. Indigenous peoples aren't often given a choice about 'progress' coming to their section of the rainforest. As a result, tension can be created between these indigenous communities and the government. In 1999, the Bakun Dam Project began in Malaysia, resulting in the eviction of approximately 10 000 indigenous people from their ancestral homeland. While they were resettled as compensation, the land provided was too small to support their traditional forms of hunting and agriculture and many failed to adapt to their new lifestyles.

Landslides

A landslide, the downward movement of earth and rocks on a slope, occurs in the lithosphere (see subtopic 6.2). It can be caused by natural physical processes such as rainfall and earthquakes, or by man-made activities such as deforestation and road building. Usually, the roots of rainforest plants keep the soil together and add stability to mountainous areas. This is especially important during times of heavier rainfall. However, sometimes the ground becomes so waterlogged that the roots can't keep the soil in place and it slips downhill, creating a landslide. The risk of this increases if deforestation has taken place on the hillside, as there are no tree roots to provide added stability.

Therefore, when these hills are cleared and settled by communities, the danger of property damage, and even death, increases. November 2011 saw 35 people killed in a landslide in the Colombian city of Manizales. Fourteen houses were destroyed, displacing up to 159 people. This mountainous, coffee-growing region used to be rainforest before it was cleared and settled.

Haiti is at a high risk of landslides because its people cut down trees to use as fuel. As a result, most of Haiti's natural forest has been destroyed. In 2004, Hurricane Jean hit the island; many of the 3000 people who died were caught in landslides.

Disease

The arrival of new tropical diseases is a less obvious result of deforestation. As animal **hosts** disappear and new human settlers move into previously inaccessible areas, 'new' disease-causing microorganisms are transferred into the human population. The frequency of mosquito-borne diseases such as malaria has increased due to the creation of more water puddles, for example in ditches and tyre treads, that are an excellent breeding ground for the mosquito. It is

FIGURE 6 A forested hillside (a) before and (b) after deforestation

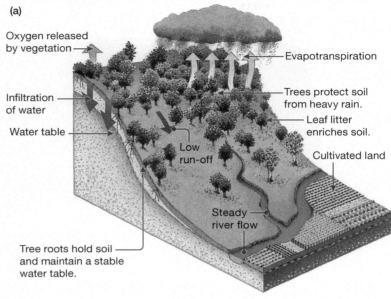

(a)

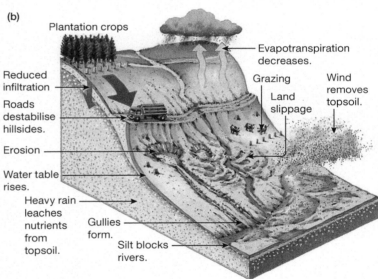

(b)

FIGURE 7 Landslides in intact forest in the Mata Atlantica rainforest, Brazil

FIGURE 8 Landslide in Manizales, Colombia, in November 2011

FIGURE 9 The deforestation in Haiti is clearly evident on its border with the Dominican Republic.

estimated that malaria is responsible for the deaths of 20 per cent of the Yanomami people in Brazil and Venezuela. Today, more than 99 per cent of malaria cases in Brazil occur in the Amazon Basin region, even though the mosquitoes that carry the disease are found across 80 per cent of the country.

The outbreak of such diseases doesn't affect only the local area but the impact can also spread into other countries via people who visit these areas, unknowingly contract an illness and then travel home, spreading the disease along the way.

6.10 Activities

To answer questions online and to receive **immediate feedback** and **sample responses** for every question, go to your learnON title at www.jacplus.com.au. *Note:* Question numbers may vary slightly.

Remember

1. Name some species threatened by deforestation in Indonesia.
2. List the main threats to orangutans.
3. What is the *interconnection* between deforestation and the impact of disease on indigenous peoples?

Explain

4. How does deforestation affect the lithosphere, atmosphere and biosphere? (Refer to subtopic 6.2 to refresh your memory.)
5. Why does having separate small islands of vegetation make it more difficult for animals to communicate and breed?

Discover

6. Refer to figures 4 and 5. Describe the *interconnection* between the two sets of data.
7. Refer to figure 6. Write a paragraph that explains how deforestation results in the consequences and *changes* illustrated in the diagram.
8. Research and create a list of 10 other animal species threatened by deforestation around the world. Choose one of these animals and report back to the class on its current location, the remaining population level and the main causes of deforestation. Present your report as a poster, PowerPoint presentation, movie (documentary), poem, song or drama performance.
9. Using the internet, investigate two different management strategies/policies/laws that have been implemented around the world to try to conserve the rainforest *environment*. Note the positive and negative aspects of these strategies. Comment on their ability to support the *sustainable* use of rainforests. Discuss your results as a class. Create a summary on the board to evaluate all the options that are shared.

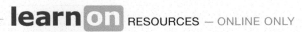

learn **on** RESOURCES — ONLINE ONLY

Try out this interactivity: Deforestation dilemma (int-3113)

6.11 How can rainforests be conserved?

6.11.1 Options for conserving rainforests

As people begin to realise the importance of rainforests, many have started to work towards preserving these valuable 'green dinosaurs'. Some methods of conservation are relevant only to governments and large companies, but some are relevant to you and the choices you make.

6.11.2 Rescue package 1: protect the remaining rainforests

While only six per cent of the world's rainforests are in a national park or reserve, there are many large areas of rainforest under protection. The number and size of these national parks are slowly increasing. The Korup National Park in Cameroon holds 126 000 hectares of Africa's richest untouched rainforest; the Khao Yai National Park in Thailand has 200 000 hectares, where the habitats of tigers, elephants and gibbons are protected; Costa Rica's rainforests are the most protected of all, with national parks and reserves covering almost one-third of that country.

6.11.3 Rescue package 2: use the forest without destroying it

This is called **sustainable development**. It means that resources are taken from the rainforests but the forest remains largely intact. It has been estimated that a forest used this way is worth $12 000 a hectare, while it is worth only $300 a hectare if it is cleared for farming.

Timber users can now purchase timber from forests that are properly managed. A company in Mexico — the Forest Stewardship Council (FSC) — assesses forests around the world. If the forests comply with regulations, the timber

FIGURE 1 The drill, one of Africa's most endangered primates, has a safe haven in the Korup National Park in Cameroon.

is given the FSC stamp. People who purchase this timber know that the forest it came from is being responsibly managed.

6.11.4 Rescue package 3: use alternative timber

One further step is not to use rainforest timber at all. Many rainforest trees are now grown in plantations, and alternatives such as using steel beams in houses and recycled paper in cardboard help take the strain off the rainforests.

One alternative that has been developed is the processing of old coconut palms to create hardwood. The company that is developing this resource, Tangaloa, claims that there are enough non-productive coconut palms to produce timber equivalent to one million rainforest trees. If this concept proves popular, plantations of coconut palms could be grown specifically for this purpose.

6.11.5 Rescue package 4: act now!

While most of us do not have rainforests growing in our backyards, the choices we make each day can and do make a difference to the way resources are used around the world. There are many organisations that aim to conserve the world's remaining rainforests. Some of their suggestions are:

- use less wood and paper
- write to businesses that destroy the rainforest
- educate yourself about the importance of rainforests
- look for alternatives to rainforest products
- be an **ecotourist** — visit rainforests where your tourist dollars go towards education and conservation.

TABLE 1 Countries with FSC-certified forests totalling more than one million hectares (2016)

Country	Area of certified forest (hectares)
Australia	1 245 429
Belarus	7 757 520
Brazil	6 176 497
Canada	52 247 475
Chile	2 353 839
China	1 142 911
Congo, The Republic of	2 766 336
Germany	1 053 684
Indonesia	2 186 470
New Zealand	1 257 931
Poland	6 933 317
Romania	2 523 283
Russia	40 752 932
South Africa	1 445 868
Sweden	12 173 649
Ukraine	2 603 965
United Kingdom	1 587 999
United States	13 876 106

FIGURE 2 Coconut plantation — could these palms help save the rainforests?

6.11 Activities

To answer questions online and to receive **immediate feedback** and **sample responses** for every question, go to your learnON title at www.jacplus.com.au. *Note:* Question numbers may vary slightly.

Remember

1. What percentage of the world's rainforests are in national parks or reserves?
2. Which country has the most protected rainforests?
3. How are rainforest *environments* in Costa Rica protected?
4. Explain in your own words what the FSC does to help protect the rainforest *environment*.

Explain

5. List two advantages and two disadvantages of each rescue package. Which of the four packages do you think offers the most hope for rainforest conservation and *sustainability*? Explain why.
6. Why is it good to have a variety of action options?

Discover

7. On a countries outline map of the world, shade in those countries with FSC-certified forests of over one million hectares. Use lighter shades of one colour for countries with smaller areas of certified forest (such as 1 000 000 – 2 499 999 and 2 500 000 – 4 999 999 hectares), and darker shades of the same colour for countries with larger areas (5 000 000 – 7 499 999; 7 500 000 – 9 999 999; >10 000 000 hectares). This type of map is called a choropleth map.
8. Other methods to help conserve the world's rainforests include:
   • breeding endangered rainforest animals in captivity, and then releasing them
   • providing websites where sponsors can give money to buy some rainforest and put it into a reserve
   • employing indigenous people to pick nuts and berries or even to breed butterflies for collectors.
   Use the internet to find an example of each of these methods and list any others that you find while completing this research. Document your findings.
9. Design your own website encouraging people to donate money to save the rainforest *environment*.

 RESOURCES — ONLINE ONLY

 Try out this interactivity: Protecting or plundering rainforests (int-3114)

6.12 Review

6.12.1 Review

The Review section contains a range of different questions and activities to help you revise and recall what you have learned, especially prior to a topic test.

6.12.2 Reflect

The Reflect section provides you with an opportunity to apply and extend your learning.

Access this subtopic at **www.jacplus.com.au**

TOPIC 7
Fieldwork inquiry: How does a waterway change from source to sea?

7.1 Overview

Numerous **videos** and **interactivities** are embedded just where you need them, at the point of learning, in your learnON title at www.jacplus.com.au. They will help you to learn the content and concepts covered in this topic.

7.1.1 Scenario and your task

Everybody lives in a catchment and its health is influenced by the activities in all areas within it. Your local water authority has received contradictory reports about the current state and health of your local catchment. As the reports are contradictory and the local water authority is not sure which is valid and which is not, they need to undertake a detailed study of the natural and built environments in the local catchment area. This will put them in an expert position to question and quash statements made by non-experts.

Your task

Your team has been commissioned by the local water authority to compile and present a report evaluating the current state of your local catchment. Your team must gather data to investigate how the catchment changes from the upper reaches to the lower. Your investigations will cover river characteristics such as depth, width and other channel characteristics, the fauna and flora in the area, and the land use in the catchment. In order to ensure that your report is accurate, your team can gather data about a local waterway and its immediate catchment by observing, collecting, interpreting and presenting your findings.

7.2 Process

7.2.1 Process

- You can complete this pro-ject individually or invite members of your class to form a group.
- **Planning:** You will need to research the characteristics of your local catchment area. In order to complete sufficient research, you will need to visit a number of sites within the catchment, comparing

different locations upstream and downstream of one creek or river. Access research topics in the Resources tab to provide a framework for your research:

- **What** sort of data and information will you need to collect at your fieldwork sites?
- **How** will you collect and record this information?
- **Where** would be the best locations to obtain data? You can determine this once you know which waterway(s) you are visiting.
- **How** will you record the information you are collecting? Consider using GPS, video recorders, cameras and mobile devices (laptop computer, tablet, mobile phone).

• Before going out into the field, examine topographic maps and aerial photos or satellite images of the relevant area to identify key landmarks (such as the location of your school, and the location of the waterway relative to the school). Locate the catchment boundary, the path of the waterway and the watercourse it contributes to. Construct a sketch map of the waterway — this map should show the catchment boundary/watershed, the river channel and the direction in which the water is flowing. Clearly note compass directions on the map. Gather spatial (mapped data) information about the region (using, for example, street directories, topographic maps, aerial photos and satellite images from sources such

as Google Earth) and information about planning, population, land use, and flora and fauna.

• Discuss with your group what you might already know about your catchment and then divide the research tasks between you. Discuss the information you will be looking for and where you might find it. Choose land use categories that you will be able to recognise and a mapping symbol to be used for each. You can view and comment on other group members' articles and rate the information they have entered.

7.2.2 Collecting and recording data

Depending on the catchment you visit, you could investigate some or all of the following:

• channel depth at various points across the stream
• channel width
• channel cross-section
• stream flow velocity (how fast the water is flowing)
• flora transects
• fauna surveys
• land-use surveys.
  Other relevant observations may include:
• condition of the waterway banks
• general slope
• native and exotic vegetation
• cleared land
• evidence of erosion
• land-use zones
• potential pollution sources (including stormwater drains entering the waterway and sewage overflow points)

- building sites, industrial and residential areas
- pollution control devices
- erosion control.

Ensure that you take relevant measuring equipment into the field, and that several measurements are taken at each site. It is useful to divide tasks among groups and then share data when you are back at school. Use a copy of your map to record the information at each site.

7.2.3 Analysing your information and data

Once you have collected, collated and shared your data, you will need to decide what information to include in your report and the most appropriate way to show your findings. If using spreadsheet data, make total and percentage calculations. Some measurements are best presented in a table, others in graphs or on maps. If you have used a spreadsheet, you may like to produce your graphs electronically. Use photographs as map annotations (either scanned and attached to your electronic map or attached to your hand-drawn map) to show features recorded at each site. You may also like to annotate each photograph to show the geographical features you observed. Describing and interpreting your data is important.

Access the report template and the presentation planning template from the Resources tab to help you complete this project. Use the report template to create your report. Use the presentation template to create an engaging presentation that showcases all of your important findings.

7.2.4 Communicating your findings

You will now produce a fieldwork report and presentation of your findings. Your report should include all of the research that you completed and all evidence to support your findings. Ensure that your report includes a title, an aim, a hypothesis (what you think you will find, which is written before you go into the field), your findings and a conclusion. You will also need to recommend some type of action that needs to be taken to improve river management at the creek or river you visited.

7.3 Review

7.3.1 Reflecting on your work

Think back over how well you worked with your group on the various tasks for this inquiry. Determine strengths and weaknesses and recommend changes you would make if you were to repeat the exercise. Identify one area where you were pleased with your performance, and an area where you would like to improve. Write two sentences outlining how you might be able to do this.

Print your Research Report and hand it in with your fieldwork report and presentation, and reflection notes.

UNIT 2
CHANGING NATIONS

For the first time in history, more people around the world live in urban areas than in rural areas. Countries such as the United States and Australia became more urbanised during the twentieth century. In fact, Australia is now one of the most urbanised countries in the world — over 89 per cent of us live in urban areas. Many other countries, including China, Indonesia, Brazil and some countries in Africa, have experienced similar changes more recently. So what has caused our world to become so urbanised? What are the consequences of this change? How can we best manage and plan urban places for the future?

Urbanisation occurs when populations shift from rural areas to urban areas. They do this in search of work and opportunities. Australia is the most urbanised country in the world.

TOPIC 8
Urbanisation

8.1 Overview

Numerous **videos** and **interactivities** are embedded just where you need them, at the point of learning, in your learnON title at www.jacplus.com.au. They will help you to learn the content and concepts covered in this topic.

8.1.1 Introduction

There are many advantages to living in large cities — for example, the economic benefit brought about by sharing the costs of providing fresh water, electricity or other energy sources and public transport between many people. There may be social benefits, because the cities provide a wider choice of sporting, recreational and cultural events. However, there are also disadvantages of living in a large city environment.

Tokyo at night

8.2 Where do most Australians live?

8.2.1 Why do Australians live where they do?

Australians live on the smallest continent and in the sixth largest country on Earth. With a population of 23 million and an area of 7 690 000 square kilometres, our **population density** is 2.9 people per square kilometre. We may think of ourselves as an outback-loving, farming nation, but we mostly live near the coast.

Most Australians currently live within a narrow coastal strip which extends from Brisbane in the north to Adelaide in the south. Over 80 per cent of Australians live in towns that have more than 1000 residents and are located within 50 kilometres of the coast. Australians love the beach, but is it just a coastal location that can explain this uneven **population distribution** pattern?

Figure 2 shows the distribution of rainfall within Australia. Comparing figures 1 and 2, it is apparent that there is a strong interconnection between the availability of more than

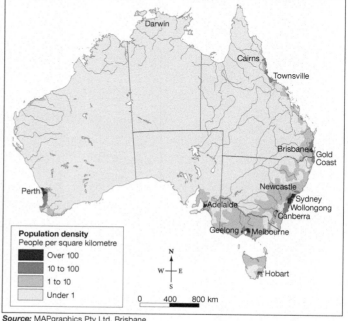

FIGURE 1 Australia's population distribution and density

Population density
People per square kilometre

- Over 100
- 10 to 100
- 1 to 10
- Under 1

Source: MAPgraphics Pty Ltd, Brisbane.

800 mm of rainfall per year and population densities of more than 10 and more than 100 people per square kilometre in the east, south-east and south-west of Australia. It would be easy to say that Australians live in places where rainfall is higher, but if you look at these maps carefully there are major exceptions to this spatial pattern. What is the relationship between population density and total rainfall in the north of Australia? Is the population density high in the regions of high rainfall in Queensland and the Northern Territory?

Coastal locations and rainfall are not the only reasons Australians live where they do. The availability of mineral resources, irrigation schemes to enhance farm production, and remote and stunning tourist destinations are **geographical factors** that draw people to live in a particular place.

FIGURE 2 The distribution of annual rainfall in Australia

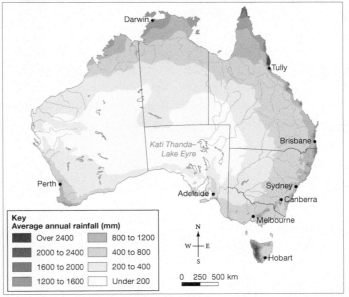

Key
Average annual rainfall (mm)

Over 2400		800 to 1200	
2000 to 2400		400 to 800	
1600 to 2000		200 to 400	
1200 to 1600		Under 200	

0   250   500 km

Source: MAPgraphics Pty Ltd, Brisbane

FIGURE 3 A remote town in northern Australia, which has a very low population density

8.2.2 How do population densities in Australia compare with those in other places?

Figure 1 shows both the population distribution and density for Australia in the present day. To better understand this data, we need to compare Australia's population density with that of other places in the world. This map shows that small areas around the major state capital cities have population densities of over 100 people per square kilometre of land. Look at table 1 and you can see that the average population density for Australia is well below the global average, and is easily the lowest of any of the permanently inhabited continents.

The population density of Australia is similar to that of Canada (3 people per square kilometre), but much lower than that of New Zealand (15 people per square kilometre), the United States (29 people per square kilometre) or China (134 people per square kilometre). Consider the geographical factors that Australia might share with Canada but not New Zealand, the United States or China that could explain the significant difference between their population densities.

TABLE 1 The average population density for each continent

Continent	Average population density (people per km^2)
Asia	95
Europe	73
Africa	34
North America	22
South America	22
Australia	3
Antarctica	0.00007

8.2 Activities

To answer questions online and to receive **immediate feedback** and **sample responses** for every question, go to your learnON title at www.jacplus.com.au. *Note*: Question numbers may vary slightly.

Remember

1. Which regions of Australia have the highest population density?
2. What is the difference between population density and population distribution?

Explain

3. What geographical factors other than rainfall may lead to the uneven distribution of population in Australia?
4. Use the statistics in table 1 to produce a world map that illustrates the contrasts between the average population densities for each continent. *Hint:* A pictograph may best highlight the differences.
5. Refer to figures 1 and 2 to produce an overlay map that identifies the *interconnection* between the distribution of population and the distribution of rainfall within Australia.
   (a) Describe areas where there are strong similarities between these two features, i.e. high population distribution and high rainfall, or low population distribution and low rainfall.
   (b) Describe *places* that have a high population distribution but low rainfall or vice versa.

Discover

6. Use your atlas to identify and list:
   (a) geographical land forms or climatic features that are common to Australia and Canada. *Hint:* Look for large regions that have an extreme climate. Explain why.
   (b) reasons New Zealand, the United States or China may have a higher population density than Australia. Explain.
7. Use various theme maps of Australia in your atlas to identify at least four possible *place or environmental* explanations for the pattern of distribution and density of Australia's population. Discuss your findings with the class.

Predict

8. Write a paragraph to explain the possible *change* in the distribution of Australia's predominantly urban population over the next 50 years if one of the following situations occurs.
   (a) The coastal urban areas become adversely affected by loss of land due to rising sea levels.
   (b) A 20-year-long drought occurs in south-eastern Australia.

Think

9. Use information from figure 2 to explain why, in the future, there may be significant movement of people from the southern states of Australia to *places* in the tropical north. Your answer must refer to specific information from the map.

8.3 Where have Australians lived in the past?

8.3.1 Before European occupation

Prior to European arrival to Australia, where did the **Indigenous** peoples live?

Until 1788, Indigenous Australian peoples inhabited all parts of Australia (see figure 1). The most densely populated areas, with 1–10 square kilometres of land per person, were the south-east, south-west and far north coastal zones, the north of Tasmania and along the major rivers of the Riverina region (south-western New South Wales).

The population density of Aboriginal and Torres Strait Islander peoples was highest in places close to coastal and river environments. These places had the best availability of food and other resources. In a location such as Port Jackson, New South Wales, food was abundant, meaning that the inhabitants needed to spend only about four hours each day hunting or gathering enough for their survival. In places where rainfall was unreliable, such as central Australia, the local peoples found it harder to survive. They often needed more than half a day to hunt and gather enough to satisfy their basic needs. When food resources ran low or with changing seasons, communities moved on to another part of their **country**. Being nomadic, they could manage their environment by not over-using the resources available at any one site.

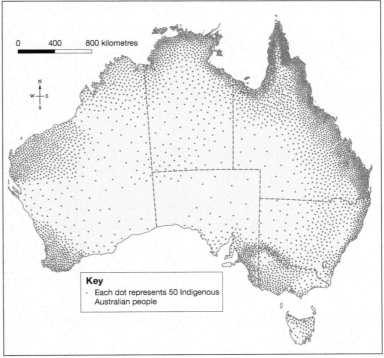

FIGURE 1 Where Aboriginal and Torres Strait Islander peoples lived in 1788

Key
· Each dot represents 50 Indigenous Australian people

Source: MAPgraphics Pty Ltd, Brisbane

8.3.2 Where do the Aboriginal and Torres Strait Islander peoples live today?

It is believed that in 1788 there were between 350 000 and 700 000 Aboriginal and Torres Strait Islander peoples, although within 50 years this population had been greatly reduced by disease and British colonists. There are currently more than 520 000 Aboriginal and Torres Strait Islander peoples, making up about 2.5 per cent of Australia's population.

The Australian environment has changed significantly since 1788. Much land has been cleared, shaped and blasted for cities, farms and mines. Other than the management of vegetation by fire, prior to European colonisation the landscape of Australia had not been greatly altered by its human inhabitants. By the twenty-first century, little of Australia's environment has not been changed significantly by human occupation.

The patterns in figures 1 and 2, showing the distribution of Aboriginal and Torres Strait Islander populations in 1788 and 2006, are generally very similar. Since before 1788, most of Australia's peoples have tended to live in the same relatively small region of this country.

FIGURE 2 Where Aboriginal and Torres Strait Islander peoples live today

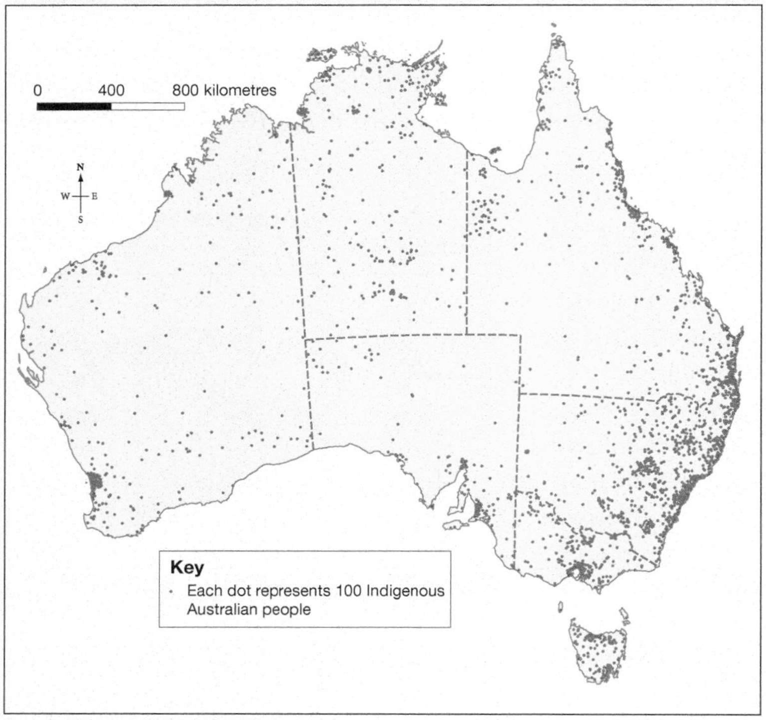

0   400   800 kilometres

Key
· Each dot represents 100 Indigenous Australian people

Source: MAPgraphics Pty Ltd, Brisbane

FIGURE 3 Many Aboriginal and Torres Strait Islander families enjoy living in remote parts of the country.

FIGURE 4 Regional distribution of Aboriginal and Torres Strait Islander peoples and the non-Aboriginal and Torres Strait Islander population of Australia

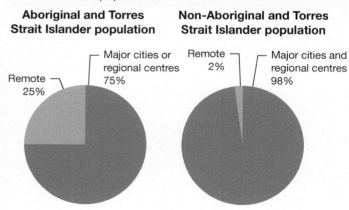

Aboriginal and Torres Strait Islander population

Remote 25%

Major cities or regional centres 75%

Non-Aboriginal and Torres Strait Islander population

Remote 2%

Major cities and regional centres 98%

FIGURE 5 The eight main climatic zones of Australia

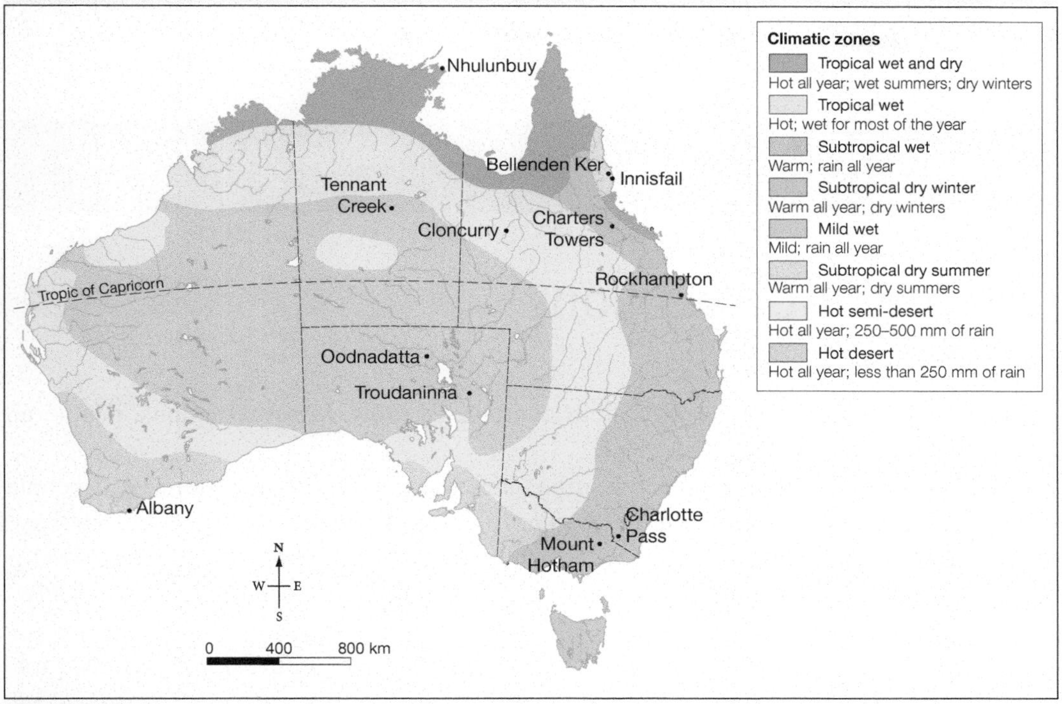

Climatic zones

- Tropical wet and dry
  Hot all year; wet summers; dry winters
- Tropical wet
  Hot; wet for most of the year
- Subtropical wet
  Warm; rain all year
- Subtropical dry winter
  Warm all year; dry winters
- Mild wet
  Mild; rain all year
- Subtropical dry summer
  Warm all year; dry summers
- Hot semi-desert
  Hot all year; 250–500 mm of rain
- Hot desert
  Hot all year; less than 250 mm of rain

Source: Spatial Vision

8.3 Activities

To answer questions online and to receive **immediate feedback** and **sample responses** for every question, go to your learnON title at www.jacplus.com.au. *Note:* Question numbers may vary slightly.

Remember

1. How many Aboriginal and Torres Strait Islander peoples:
   (a) lived in Australia in 1788
   (b) live in Australia today?

Explain

2. (a) Identify the climatic zones in figure 5 that best match the population density areas in figure 1.
   (b) For each of the states shown in figure 1, write a sentence to describe the climate for the region. For example, 'This region has a mostly mild to subtropical climate with rainfall all year round.'

Discover

3. Refer to figure 4. Living so far away from major cities means that 25 per cent of Aboriginal and Torres Strait Islander communities have limited access to many of the services and opportunities that cities offer their residents. In a small group, brainstorm the lifestyle and service difficulties that may be associated with living so remotely.

Think

4. Collect some statistics that identify the health, wealth and educational inequalities which exist between Aboriginal and Torres Strait Islander peoples and non-Aboriginal and Torres Strait Islanders. For example, Aboriginal males have a life expectancy 17 years less than that of non-Aboriginal males born in the same year. Use the **ABS: Indigenous health** weblink in the Resources tab to start your research. Write a paragraph or produce a series of graphs to comment on the inequalities you have discovered.

my**World**Atlas   **Deepen your understanding of this topic with related case studies and questions.**
❂ **Indigenous Australians**

8.4 What is urbanisation?

8.4.1 Urbanisation

As the world's population increases, **urban** areas continue to grow. In some regions, people are moving from rural to urban areas at very high rates.

Urbanisation is the growth and expansion of urban areas and involves the movement of people to towns and cities. The earliest cities emerged about 5000 years ago in Mesopotamia (part of present-day Iran, Iraq and Syria). Originally these cities depended on agriculture. In 1800, 98 per cent of the global population lived in rural areas and most were still dependent upon farming and livestock production — only 2 per cent of people lived in urban areas.

However, as cities grew and trade developed, urban areas became centres for merchants, traders, government officials and craftspeople. By 2008, the number of people living in urban areas had increased to 50.1 per cent, and in 2016 the figure had risen again to 54.5 per cent (see also figure 1). The rate of growth has varied in different regions (see figure 2).

FIGURE 1 Percentage of population living in urban centres, 2014

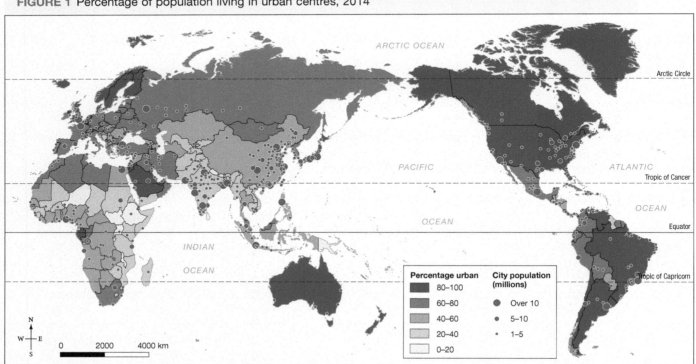

Source: United Nations, Department of Economic and Social Affairs, Population Division (2014). World Urbanization Prospects: The 2014 Revision, CD-ROM Edition.

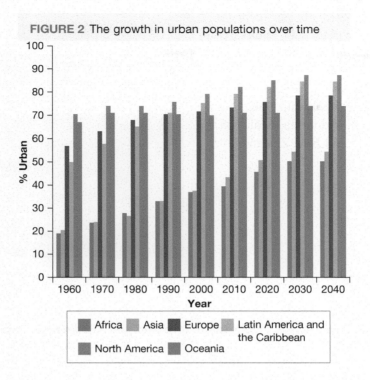

FIGURE 2 The growth in urban populations over time

% Urban (y-axis, 0 to 100)

Year (x-axis): 1960, 1970, 1980, 1990, 2000, 2010, 2020, 2030, 2040

Legend: ■ Africa ■ Asia ■ Europe ■ Latin America and the Caribbean ■ North America ■ Oceania

8.4.2 Uneven urbanisation

Urbanisation has not occurred evenly across the world. Some countries are predominantly rural, such as Cambodia and Papua New Guinea (populations 79 per cent and 87 per cent rural respectively), whereas others are almost completely urban, such as Belgium and Kuwait (98 per cent urban for both). In fact, some countries have 100 per cent urbanisation, including Bermuda, Cayman Islands, Hong Kong, Macau, Monaco and Singapore. South America is becoming one of the most urbanised regions in the world and currently has a population of around 385 million people. It is estimated that, by 2050, 91.4 per cent of its population will be residing in urban areas.

FIGURE 3 Urban housing in Kuwait

Coastal urbanisation

People have lived on coastlines for thousands of years. Often at the mouth of rivers, coastal settlements became centres of trade and commerce and quickly grew into cities. Today, about half the world's population lives along or within 200 kilometres of a coastline (see figure 4). This means about 3.2 billion people live on only 10 per cent of the Earth's land area.

Countries that have over 80 per cent of their population living within 100 kilometres of a coastline include the United Kingdom, Senegal, Portugal, Belgium, the Netherlands, Sweden, Norway, Tunisia, Greece, Oman, the United Arab Emirates, Kuwait, Qatar, Sri Lanka, Japan, Singapore, Indonesia, Malaysia, the Philippines, Australia and New Zealand.

FIGURE 4 Cape Town in South Africa is a city located on the coast.

8.4 Activities

To answer questions online and to receive **immediate feedback** and **sample responses** for every question, go to your learnON title at www.jacplus.com.au. *Note*: Question numbers may vary slightly.

Remember

1. Define *urbanisation* in your own words.

Explain

2. How has urbanisation changed from 1950 to the present? How is this different around the world? What is expected to happen in the future?
3. Explain how figure 2 shows that urbanisation has varied in different regions of the world. Which two regions have the greatest rural population?
4. Use the **Urbanisation Gapminder** weblink in the Resources tab to watch an animation on urbanisation.
   (a) What does the graph show?
   (b) What do the colours represent?
   (c) In 1963, name two countries that were at the bottom of the graph and two at the top.
   (d) Why does Singapore appear where it does on the graph? What is an 'urban nation'?
   (e) Why do the 'bubbles' increase in size over time?
   (f) Which regions were highly urbanised in 2004?
   (g) What do the positions of the bubbles show about urbanisation in Africa, South Asia and China?

Discover

5. Refer to a world population density map in your atlas or online. Compare this map with the two regions that have the highest rural population. What pattern do you see?
6. Look at figure 1, which shows the population in urban areas. Identify and name the three countries with the highest and the three with the lowest percentage of people living in urban areas. Write a description of the general pattern shown in the map. Include patterns within different continents in your description.

Predict

7. Look at a physical map in an atlas to locate the countries with more than 80 per cent of their population located on the coast. Study the location of each country and create a table to record possible reasons for this pattern.

Think

8. Rural areas are where most food is produced. What are two possible outcomes for food production if urbanisation continues?
9. Draw a sketch of the photograph of Cape Town in figure 4. Annotate the sketch, identifying the possible advantages and disadvantages to the natural environment when cities and towns are located on the coast.

 my**World**Atlas **Deepen your understanding of this topic with related case studies and questions.**
➲ World urbanisation

8.5 Is Australia an urbanised country?

8.5.1 Australians in urban areas

With a population of 24 million people and a very large landmass, Australia has an average population density of only 3.1 people per square kilometre. Yet 84 per cent of people live within 50 kilometres of the coast, and most of these people — in 2014, 89 per cent of Australians — live in urban areas.

Australia is one of the most urbanised and coast-dwelling populations in the world and the level of urbanisation is increasing (see figure 1). From Federation (1901) until 1976, the number of Australians living in capital cities increased gradually from a little over one-third (36 per cent) to almost two-thirds (65 per cent). From 1977 to the present, the population in capital cities has grown to 66 per cent. It is estimated that by 2053 this will have grown to 72 per cent (with an estimated 89 per cent in the four largest capital cities).

All Australia's capital cities have grown over time, as have many regional urban areas such as the Gold Coast and Moreton Bay regions. This growth is expected to continue in the future (see table 1).

FIGURE 1 A map of Australia's population distribution in 2011 shows that it is highly urbanised and coastal.

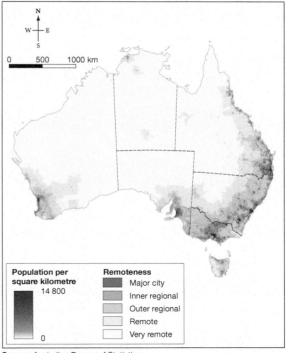

Source: Australian Bureau of Statistics

TABLE 1 Australian capital city populations 2016 and projected 2031 and 2061

City	2016 population	Projected 2031	Projected 2061
Sydney	4 986 714	6 206 843	8 493 740
Melbourne	4 605 993	5 984 219	8 580 556
Brisbane	2 397 068	3 190 129	4 787 996
Perth	2 181 194	3 248 550	5 451 406
Adelaide	1 340 503	1 566 929	1 920 727
Canberra	405 827	520 412	740 903
Hobart	222 533	247 320	270 655
Darwin	140 943	170 153	225 873
Total	16 280 775	21 134 555	30 471 856

8.5.2 What are the consequences of a highly urbanised Australia?

More land is needed when cities expand and this results in the greatest change — from agricultural to urban land. This has been called **urban sprawl**. Sydney's Greater Metropolitan Region now extends from Port Stephens in the north to Kiama in the south. Some townships in the Blue Mountains are now also considered part of Sydney despite being located 50–120 kilometres west of Sydney's CBD (see figure 2). Melbourne, Perth and Brisbane have also spread into distant, previously agricultural areas.

Historically, urban areas were settled where the land was flat, the water and soil were good and the climate was temperate — in other words, where good farmland is located. When cities spread, the sprawl takes over arable land (land able to be farmed for crops). Urban sprawl has long-term effects, as it is very difficult to bring the soil back to its former state once the predominant land use has been for buildings.

Many of Australia's cities have been called 'car cities' due to the reliance on cars and road networks for transport. These have an impact on commuting times to and from workplaces (see figure 3).

8.5.3 Ecological footprint

The amount of productive land needed on average by each person (in the world or in a country, city or suburb, for example) for food, water, transport, housing and waste management is known as an **ecological footprint**. It is measured in hectares per person per year. The average global ecological footprint is 2.84; Australia has an average ecological footprint of 9.3 (see table 2). The country with the highest ecological footprint is Luxembourg with 15.8.

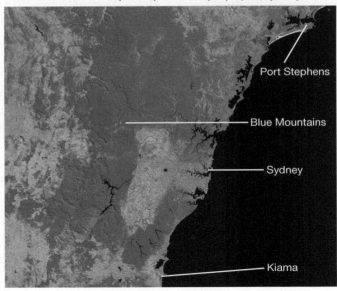

FIGURE 2 Urban sprawl (shown in purple) in Sydney

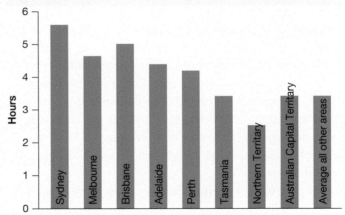

FIGURE 3 Average weekly commuting times in selected Australian cities

Source: AMP/NATSEM Race against time report, 2011

TABLE 2 Ecological footprints of Australian capital cities

City	Ecological footprint value (hectares/person/year)
Perth	7.66
Canberra	7.09
Darwin	7.06
Brisbane	6.87
Sydney	6.82
Adelaide	6.72
Melbourne	6.33
Hobart	5.50

8.5 Activities

To answer questions online and to receive **immediate feedback** and **sample responses** for every question, go to your learnON title at www.jacplus.com.au. *Note*: Question numbers may vary slightly.

Remember

1. What percentage of Australians live in urban areas? Of these, what percentage live in urban areas close to the coast?
2. List the disadvantages of urban sprawl.

Explain

3. Refer to figure 1 and describe the population distribution of Australia.
4. Refer to table 1. Draw a bar graph to show the predicted *change* in the populations of Australia's capital cities. What does your graph reveal?
5. Describe the spread of Sydney shown in figure 2. Use an atlas to explain why growth has not occurred as much to the north and south as it has to the west.
6. Describe the average commuting times shown in figure 3. What reasons can you give for the differences between cities?

Discover

7. Conduct research to find which country in the world has the highest average population density. Find one country with a lower average population density than Australia.

Predict

8. Use your atlas or online research to find an urban growth map for the capital city in your state or territory. Describe the *change* that has taken place over time. Using this map and a physical map of your state or territory, predict where future growth might occur. Justify your responses.

Think

9. (a) What is an ecological footprint?
   (b) Refer to table 2. How does the ecological footprint data compare for Australian cities?
   (c) How do these figures compare with the average global ecological footprint?
   (d) Use the **UAE ecological footprint** weblink in the Resources tab to watch a video. How does the ecological footprint in the United Arab Emirates compare to that of Australian cities? What would happen if all cities had such a high footprint?
   (e) Create your own advertisement or animation using a video editing program to encourage people in your capital city to reduce their ecological footprint.
10. Study table 2 and your graph again. Consider the issues and problems that increasing city populations will create. Discuss this as a class and construct a consequence chart to summarise all the ideas. What might be some solutions to these issues and problems? Add this to your chart.

 Explore more with this weblink: UAE ecological footprint

Deepen your understanding of this topic with related case studies and questions.
➊ **Urbanisation in Australia**

8.6 SkillBuilder: Understanding thematic maps

FIGURE 1 Thematic map of the major landform regions of Australia

WHAT IS A THEMATIC MAP?

A thematic map is a map drawn to show one aspect; that is, one theme. For example, a map may show the location of vegetation types, hazards, or weather. Parts of the theme are given different colours or, if only one idea is conveyed, symbols may show location.

Go online to access:

- a clear step-by-step explanation to help you master the skill
- a model of what you are aiming for
- a checklist of key aspects of the skill
- a series of questions to help you apply the skill and to check your understanding.`

8.7 How urban are the United States and Australia?

8.7.1 Urbanisation in the United States and Australia

Both the United States and Australia are very large countries that are highly urbanised. In fact, both are among the world's most urbanised nations.

The United States and Australia have some similarities and some differences in terms of how urbanised they are, as revealed in table 1 and figure 1.

TABLE 1 A comparison of urbanisation in the United States and Australia

	United States	Australia
Population	320 000 000 (2015 census)	23 800 000 (in 2015)
Population distribution	Over 81% live in urban areas, and 19.5% in rural areas.	Over 89% live in urban areas, less than 11% in rural areas.
People living in large cities	The United States has 10 cities that have a population of more than 1 million people.	Australia has 5 cities that have a population of more than 1 million people.
	Approximately 1 of every 10 people in the United States live in either the New York or Los Angeles metropolitan areas.	Approximately 4 of every 10 people in Australia live in either Melbourne or Sydney.

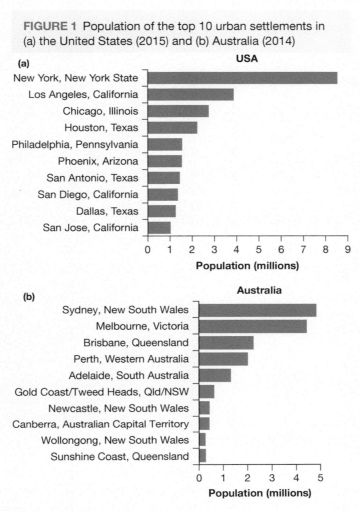

FIGURE 1 Population of the top 10 urban settlements in (a) the United States (2015) and (b) Australia (2014)

8.7.2 Causes of urbanisation

The causes of urbanisation are similar for both Australia and the United States. In each case, since the country was founded:

- fewer people were needed to work in rural areas as technology reduced the demand for labour on farms
- more jobs and opportunities were available in factories, which were located in urban areas
- the development of railways allowed goods produced in one city to be transported to rural and urban areas
- cities could grow and develop thanks to new technologies (steel-framed skyscrapers) and utilities (for example, electricity and water supply).

8.7.3 Consequences of urbanisation = conurbations

Conurbations

Sometimes there are so many cities in a particular region that they seem to merge almost into one city as they expand. A conurbation is made up of cities that have grown and merged to form one continuous urban area. Both the United States and, to a lesser extent, Australia have conurbations.

United States

Eleven conurbations have been identified in the United States (see figure 2). The major conurbation is in the north-east region. It is often called BosNYWash because it covers the area from Boston in the north, through New York to Washington in the south. This region is home to over 50 million people and accounts for 20 per cent of the gross domestic product (GDP) of the United States.

FIGURE 2 Conurbations in the United States

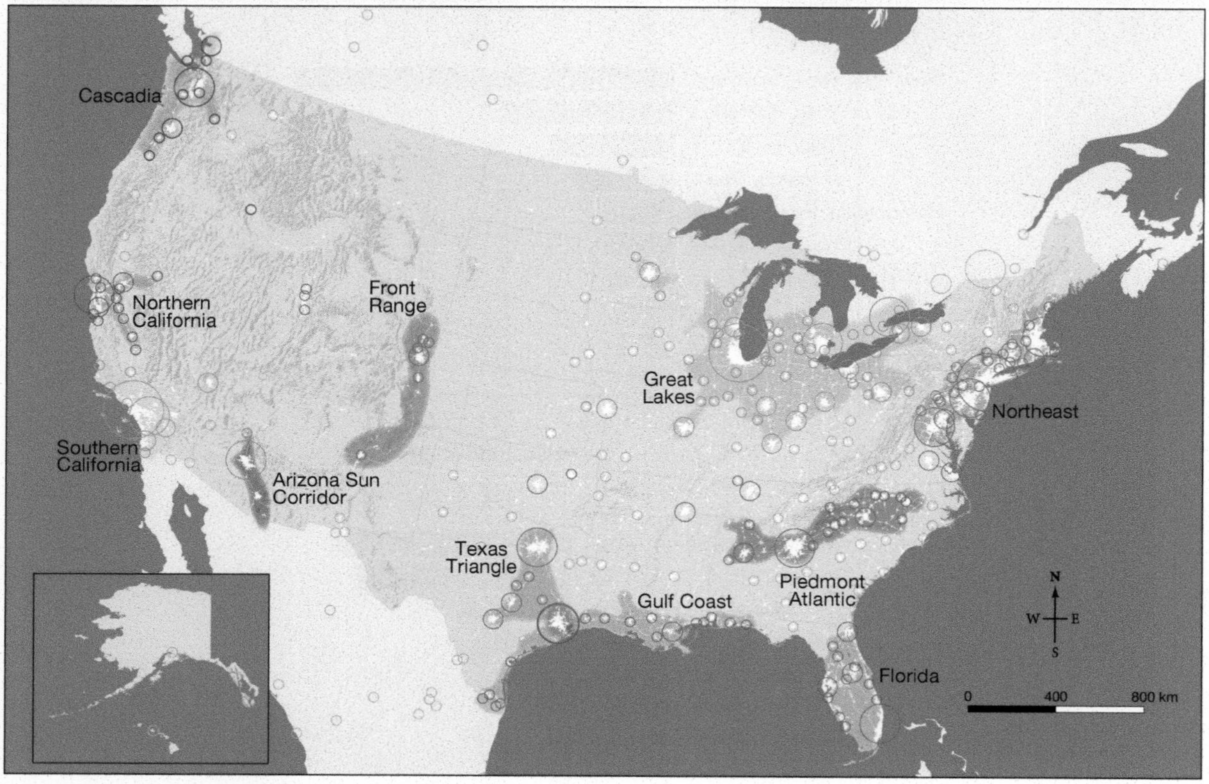

Source: Adapted with permission from Bernard Salt.

Australia

Australia, on the other hand, has only two main conurbations (see figure 3). One is in south-east Queensland; the other, the Newcastle–Wollongong conurbation, stretches for over 250 kilometres and is home to almost six million people.

8.7.4 Other consequences of urbanisation

Homelessness

It has been estimated that there are more than 560 000 homeless people (living on the streets or in temporary shelters) in the United States and more than 105 000 homeless people in Australia. One reason for this is that urban housing projects do not provide affordable housing for the poor.

In the United States the highest number of homeless per capita reside in New York (130), Atlanta (131), Boston (132), Washington (133) and Honolulu (134).

Health issues

High population densities in urban areas make it easier for diseases to be transmitted, especially in poor neighbourhoods. The urban poor suffer health issues caused by reduced access to sanitation and hygiene facilities and health care.

Pollution

Air pollution from cars, industry and heating affects people who live in cities. A study in the United States showed that more than 3800 people die prematurely in the Los Angeles Basin and San Joaquin Valley region of southern California because of air pollution. Generally, Australia has a fairly high level of air quality. Cars and industry are the main factors influencing air quality in urban areas.

FIGURE 3 Australia's population centres and conurbations

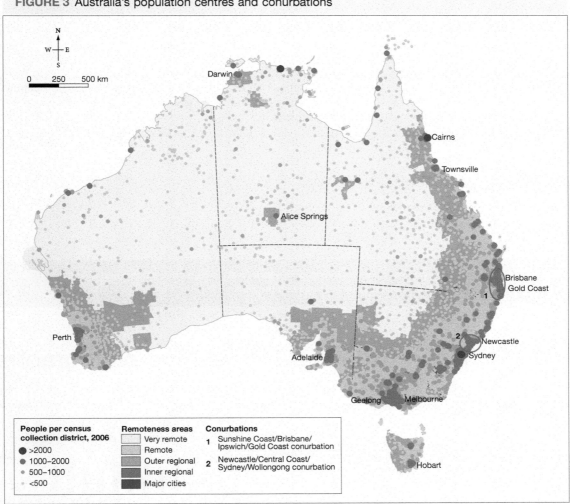

Source: Australian Bureau of Statistics

8.7 Activities

To answer questions online and to receive **immediate feedback** and **sample responses** for every question, go to your learnON title at www.jacplus.com.au. *Note*: Question numbers may vary slightly.

Remember

1. How does the population of the United States compare to that of Australia? How many times larger (approximately) is one than the other?
2. Refer to figures 2 and 3. Describe the distribution of the population in the United States and in Australia.
3. What is a conurbation?

Explain

4. Refer to table 1.
   (a) Compare the *scale* of urbanisation in the United States and in Australia.
   (b) Compare the numbers of people living in large cities in the United States and in Australia.

5. Refer to figure 1.
   (a) Compare the size of the 10 largest cities in the United States and in Australia.
   (b) What might explain the differences you noticed?
6. Explain, in your own words, the causes of urbanisation in the United States and Australia.

Think

7. Why do you think both Australia and the United States have conurbations?
8. Why might there be more conurbations in the United States than in Australia?

Discover

9. Conduct research to find out about other consequences of urbanisation in the United States and Australia, such as those affecting traffic, provision of adequate public transport, water supply and energy, waste management issues, urban sprawl and loss of farmland.

 **learn on** RESOURCES — ONLINE ONLY

 Try out this interactivity: City folk (int-3117)

my**World**Atlas **Deepen your understanding of this topic with related case studies and questions.**
 ❍ Urbanisation in Australia and the USA

8.8 SkillBuilder: Creating and reading pictographs

WHAT IS A PICTOGRAPH?

A pictograph is a graph drawn using pictures to represent numbers, instead of bars or dots which are traditionally used on graphs. A pictograph is a simple way of representing data and conveying information quickly and efficiently in a different format.

Go online to access:

- a clear step-by-step explanation to help you master the skill
- a model of what you are aiming for
- a checklist of key aspects of the skill
- a series of questions to help you apply the skill and to check your understanding.

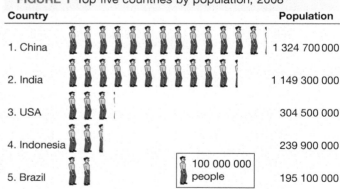

FIGURE 1 Top five countries by population, 2008

Country		Population
1. China		1 324 700 000
2. India		1 149 300 000
3. USA		304 500 000
4. Indonesia		239 900 000
5. Brazil		195 100 000

100 000 000 people

 **learn on** RESOURCES — ONLINE ONLY

 Watch this eLesson: Creating and reading pictographs (eles-1659)

Try out this interactivity: Creating and reading pictographs (int-3155)

8.9 SkillBuilder: Comparing population profiles

WHAT IS A POPULATION PROFILE?

A population profile, sometimes called a population pyramid, is a bar graph that provides information about the age and gender of a population. The shape of the population profile tells us about a particular population. Comparing population profiles of different places helps us try to understand how and why they may be similar or different.

Go online to access:

- a clear step-by-step explanation to help you master the skill
- a model of what you are aiming for
- a checklist of key aspects of the skill
- a series of questions to help you apply the skill and to check your understanding.

FIGURE 1 Population profiles of Indonesia and Vanuatu, 2010

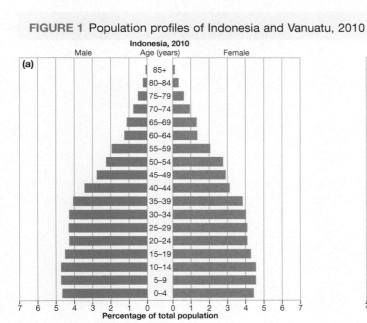
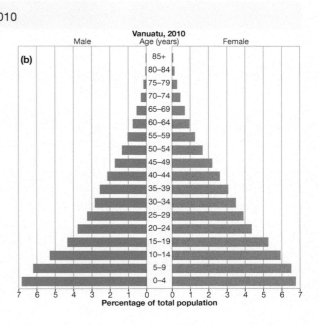

learn on ONLINE ONLY

 **Watch this eLesson:** Comparing population profiles (eles-1704)

Explore more with these weblinks: Population pyramid, US Census Bureau

Try out this interactivity: Comparing population profiles (int-3284)

8.10 How has international migration affected Australia?

8.10.1 Why have people migrated to Australia?

Australia is a land of **migrants**. In a way we are all migrants — at some stage in the past, our ancestors came to this country to live. Today, half of our population of just over 23 million people either was born overseas or has at least one parent who was born overseas.

Since the earliest times, people have moved from one part of the world to another in search of places to live. Migrants have come to Australia for many reasons (see figure 1).

FIGURE 1 Reasons for immigration to Australia

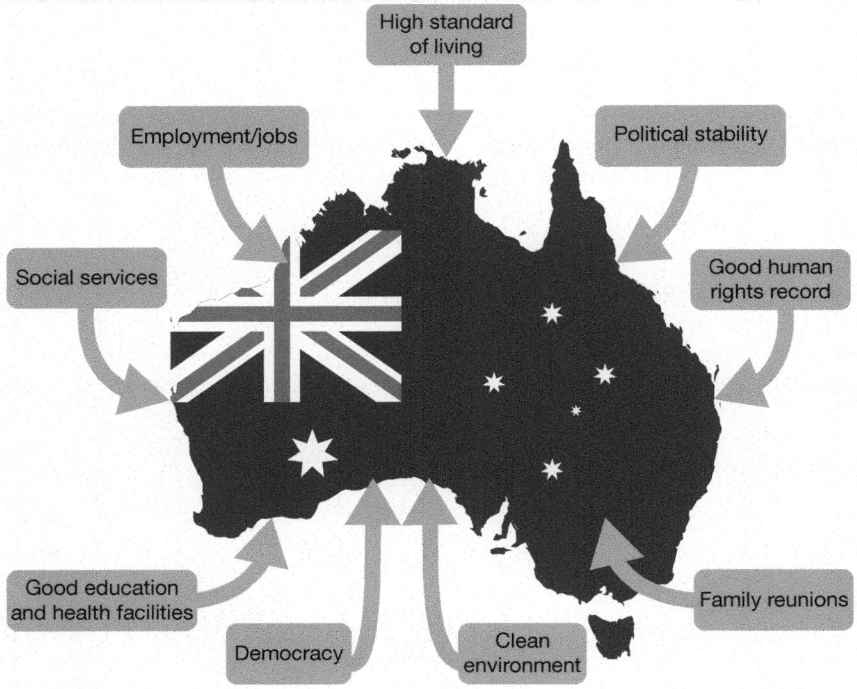

High standard of living

Employment/jobs

Political stability

Social services

Good human rights record

Good education and health facilities

Family reunions

Democracy

Clean environment

8.10.2 Where have our migrants come from?

At first, migrants to Australia came exclusively from Europe (see figure 2); however, since 1975, the country has attracted more immigrants from Asia (see figure 3 and table 2). Despite this, the most common ancestries today are still English, Australian, Irish, Scottish and Italian (see table 1).

FIGURE 2 Origin of Australia's migrants, 1949–1959

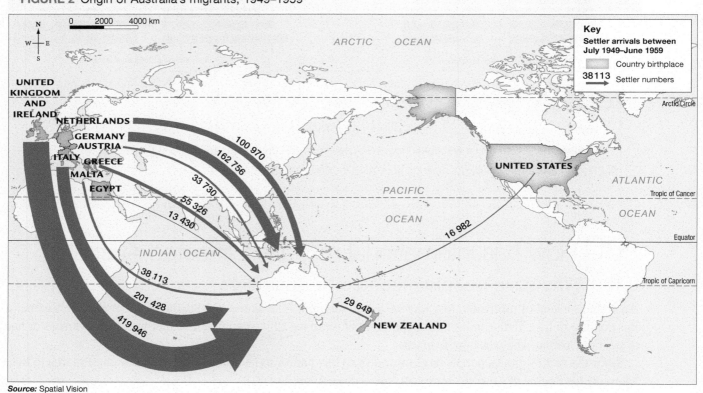

Key
Settler arrivals between July 1949–June 1959
Country birthplace
38113 Settler numbers

100 970
162 756
33 730
55 326
13 430
38 113
201 428
419 946
16 982
29 649

Source: Spatial Vision

FIGURE 3 Settler arrivals July 2012 to June 2013 by country of birth

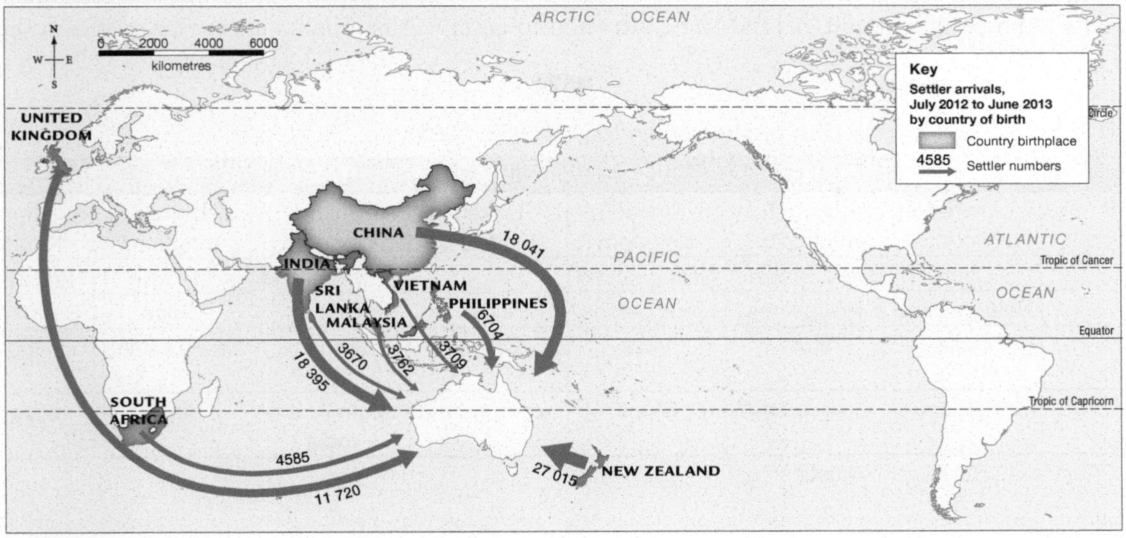

Source: Department of Immigration and Border Protection

TABLE 1 Ancestry of Australians, 2011

Ancestry (top responses)	Number of Australians	Percentage
English	7 238 533	25.9
Australian	7 098 486	25.4
Irish	2 087 758	7.5
Scottish	1 792 622	6.4
Italian	916 121	3.3

Source: © Australian Bureau of Statistics, licensed under a Creative Commons Attribution 2.5 Australia licence

Where have our migrants settled?

When they arrive, migrants tend to live in capital cities because of the greater availability of jobs and to be near family members, friends and people from the same country (see table 2). In 2011, 82 per cent of the overseas-born population in Australia lived in capital cities, compared with 66 per cent of all people in the country. About one-third of the population in our large cities was born overseas.

TABLE 2 Top 10 birthplaces of Australians, 2011

Country of birth	Number of people	Percentage of state population	Percentage of state population living in capital city
United Kingdom	1 098 638	5.1	5.6
New Zealand	483 376	2.2	2.3
China	318 965	1.5	2.1
India	295 355	1.4	1.9
Italy	185 402	0.9	1.1
Vietnam	185 032	0.9	1.3
Philippines	171 223	0.8	1.0
South Africa	145 670	0.7	0.8
Malaysia	115 790	0.5	0.8
Germany	108 001	0.5	0.5

Source: Based on Australian Bureau of Statistics data (Census of Population and Housing 2011) © Commonwealth of Australia, licensed under a Creative Commons Attribution 2.5 Australia licence

Overseas-born migrants who arrived in the past 20 years are more likely to live in a capital city than those who arrived before 1992 (85 per cent compared to 79 per cent).

Migrants from certain countries tend to be attracted to certain Australian states or territories more than others (see table 3).

TABLE 3 Top five countries of birth by state or territory, 2006 and 2011

ACT	NSW	NT	Qld	SA	TAS	VIC	WA
United Kingdom	United Kingdom	United Kingdom	United Kingdom	United Kingdom	United Kingdom	United Kingdom	United Kingdom
China	China	New Zealand	New Zealand	Italy	New Zealand	India	New Zealand
India	New Zealand	Philippines	South Africa	India	China	China	South Africa
New Zealand	India	India	India	China	Germany	New Zealand	India
Vietnam	Vietnam	Greece/USA	China	New Zealand	South Africa/India	Italy	Malaysia

Source: Australian Bureau of Statistics

For example:

- More than half of all overseas-born people in Western Australia, South Australia and Tasmania were born in the United Kingdom and Ireland. Western Australia is the state with the highest proportion of its population having been born overseas, and is home to around one in five of all British-born migrants in Australia.
- Queensland has the greatest proportion of migrants born in Papua New Guinea and New Zealand.
- Nearly half of all Australian migrants born in Greece live in Victoria. People from Sri Lanka, Turkey and Greece also tend to live in Victoria. Victoria is home to the second-largest number of overseas-born people.
- New South Wales, Victoria and Western Australia account for about 80 per cent of Asian migrants.
- Combined, Victoria and New South Wales are home to almost 80 per cent of migrants from Vietnam.

Not only have immigrants tended to settle in larger cities, they have settled in particular suburbs and regions within the capital cities. Many migrants have settled in inner Sydney, for example, and especially in western Sydney suburbs (see figure 4).

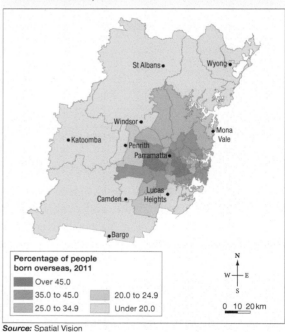

FIGURE 4 Percentage of Sydney residents born overseas, 2011

Percentage of people born overseas, 2011

Over 45.0
35.0 to 45.0 — 20.0 to 24.9
25.0 to 34.9 — Under 20.0

Source: Spatial Vision

8.10.3 Effects of international migration
Social effects

Migration has helped increase Australia's population. The increase in population from only seven million at the end of World War II to more than triple that now is caused by both the arrival of migrants and increased birth rates since then (see figure 5).

Migrants to Australia have contributed to our society, culture and prosperity. Many communities hold festivals and cultural events where we can all share and enjoy the foods, languages, music, customs, art and dance.

Australian society is made up of people from many different backgrounds and origins. We have come from more than 200 countries to live here. Therefore, we are a very multicultural society — one which needs to respect and support each other's differences, and the rights of everyone to have their own culture, language and religion.

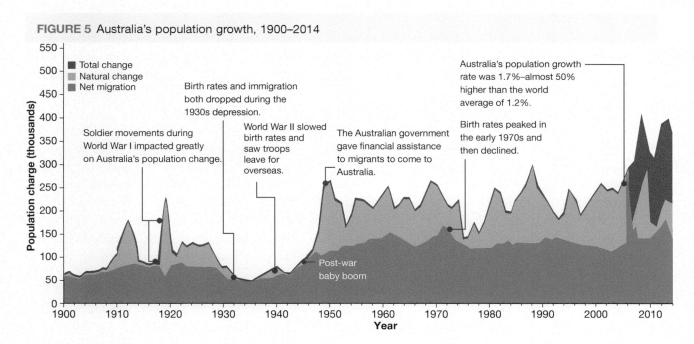

FIGURE 5 Australia's population growth, 1900–2014

Total change
Natural change
Net migration

Soldier movements during World War I impacted greatly on Australia's population change.

Birth rates and immigration both dropped during the 1930s depression.

World War II slowed birth rates and saw troops leave for overseas.

The Australian government gave financial assistance to migrants to come to Australia.

Birth rates peaked in the early 1970s and then declined.

Australia's population growth rate was 1.7%–almost 50% higher than the world average of 1.2%.

Post-war baby boom

Economic effects

An increased population also means a greater demand for goods and services, which stimulates the economy. Migrants need food, housing, education and health services, and their taxes and spending allows businesses to expand. Apart from labour and capital (money), migrants also bring many skills to Australia (see figure 6).

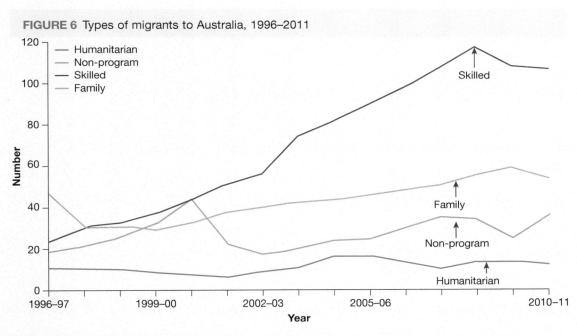

FIGURE 6 Types of migrants to Australia, 1996–2011

Humanitarian
Non-program
Skilled
Family

Source: © G. Hugo, *Australia's changing population and the future*, presentation to the Migration Institute of Australia Migration 2010 conference, Sydney, 8 October 2010, and advice provided by the author to the Parliamentary Library, March 2012. Data sources: ABS 2007, *Australian Social Trends*, DIAC 2009 and 2011

Migrants generate more in taxes than they consume in benefits and government goods and services. As a result, migrants as a whole contribute more financially than they take from society.

Environmental effects

In the past, people argued that immigrants put pressures on Australia's environment and resources by increasing our population and the need for water, energy and other requirements. Today, however, many people believe that Australia's environmental problems are not caused by migration and population increase, but by inadequate planning and management.

8.10.4 The future

Since 1995, the Australian government has been working to encourage new migrants to settle in regional and rural Australia. The Regional Sponsored Migration Scheme (RSMS) allows employers in areas of Australia that are regional, remote or have low population growth to sponsor employees to work with them in those regions (see figure 7). This takes the pressure off large cities and also provides regional employers with skilled workers. As we have seen, it has always been the case that most immigrants settle first in our cities, especially the state capitals. However, more migrants are now choosing to settle initially in regional areas. Around one in six new permanent arrivals are settling in regional Australia. There are many regional locations that want to attract migrants.

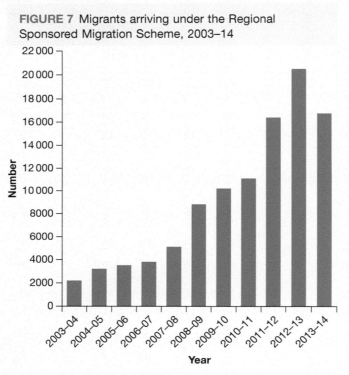

FIGURE 7 Migrants arriving under the Regional Sponsored Migration Scheme, 2003–14

Source: © Department of Immigration and Citizenship, *2011–12 Migration Program Report*

8.10 Activities

To answer questions online and to receive **immediate feedback** and **sample responses** for every question, go to your learnON title at www.jacplus.com.au. *Note*: Question numbers may vary slightly.

Remember

1. Using statistics, describe how Australia is truly a land of migrants.
2. Refer to figures 2 and 3. Describe how the origins of our migrants have *changed* since 1949.

3. Refer to figure 7. Describe how the number of migrants coming into Australia under the Regional Sponsored Migration Scheme has *changed* between 2003–04 and 2011–12.
4. Refer to table 3 and figure 4. Describe how the distribution of the areas of settlement by migrants varies within Australia.

Think

5. Refer to figure 5. Describe how important migration has been in terms of Australia's population growth.
6. What do you consider to be the main reasons for why people would migrate to Australia?
7. What do you believe are the two main benefits of migration to Australia? Give reasons for your answer.

Deepen your understanding of this topic with related case studies and questions.
❍ International migration and Australian cities

8.11 Why are people on the move in Australia?

8.11.1 What makes Australians move?

In the United States, it is common for young people to leave home and travel to a university in another state or on the opposite side of the country. This is less common in Australia.

People move for many reasons. The average Australian will live in 11 houses during their lifetime — this means that many people will live in more. You may move to live in a larger house, or a smaller house as your family size or income changes. On retirement you may want to live near the mountains or the sea.

Forty per cent of Australians changed the place where they live in the five years between 2006 and 2011 (see table 1). Some moved only within their suburb, but 6 per cent, or 1.2 million people, moved from a different country.

The major movements of Australians since 1788 are shown in figure 1. The Great Australian Divide separates Australia into two regions, known as the Heartland and the Frontier. The Heartland is home to about 19 million people who live in a modern, urbanised, industrial state. The Frontier is a sparsely populated region of only about three million people who live in a place that is remote but rich in resources.

TABLE 1 The usual place of residence for Australians in 2006 versus 2011

Location	Number of people
Same address	11 million
Same suburb, different address	1.4 million
Different suburb or state	5.2 million
Overseas	1.2 million
Not stated	1.2 million
Total	20 million

Sea change or tree change

The population movement caused by 'sea change' or 'tree change' — a move from an urban environment to a rural location — is a national issue affecting coastal and forested mountain communities in every state in Australia. The movement involves people who are searching for a more peaceful or meaningful existence, who want to know their neighbours and have plenty of time to relax. Local communities in high-growth coastal and mountain areas often cannot afford the services and increased infrastructure, such as roads, water and sewerage, that a larger population requires. Geelong, Wollongong, Cairns and the Gold Coast are all popular places for sea changers to settle.

Not every sea changer loves their new life, and many return to the city. Factors such as distance from family, friends, cultural activities and various professional or health services may pull people back to their previous city residences.

FIGURE 1 Australia's moving population

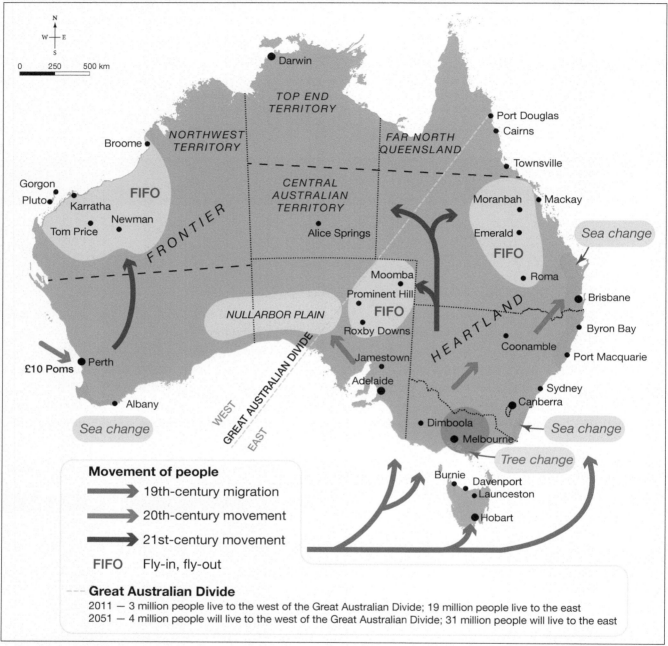

Movement of people

→ 19th-century migration

→ 20th-century movement

→ 21st-century movement

FIFO Fly-in, fly-out

---- **Great Australian Divide**

2011 — 3 million people live to the west of the Great Australian Divide; 19 million people live to the east

2051 — 4 million people will live to the west of the Great Australian Divide; 31 million people will live to the east

Source: MAPgraphics Pty Ltd, Brisbane

8.11.2 Fly-in, fly-out workers

Employment opportunities have grown within the mining industry in places such as the Pilbara. However, local towns do not have the infrastructure, such as water, power and other services, to support a large population increase. Rental payments for homes can be as high as $3000 per week; many places do not have mobile phone reception, and even internet access can be costly and slow. One way to attract workers to these regions is to have a **fly-in, fly-out (FIFO)** workforce. FIFO workers are not actually 'settlers', as they choose not to live where they work. Some mine workers from the Pilbara live in Perth or even Bali, and commute to their workplace on a weekly, fortnightly or longer-term basis. The permanent residents of these remote towns are uneasy with the effects of the FIFO workforce because they change the nature of the town but choose not to make it their home. By not living locally, their wages leave the region and are not invested in local businesses and services.

8.11.3 Seasonal agricultural workers

Many jobs in rural areas are seasonal — for example, the picking and pruning of grapes and fruit trees requires a large workforce for only a few months each year. Many children born in rural areas leave their homes and move to the city for education, employment or a more exciting lifestyle than the one they knew in the country. This means that there are not enough agricultural workers to cover the seasonal activities.

Backpackers plus people from Asia and the Pacific Islands on short-term work visas often provide the seasonal workforce in these regions. Country towns such as Robinvale in northern Victoria now have Asian grocery stores, an Asian bakery and a shop selling Tongan canned goods, providing the seasonal farm work-force with a taste of home. Robinvale has more than 20 nationalities as either residents or seasonal workers.

8.11 Activities

To answer questions online and to receive **immediate feedback** and **sample responses** for every question, go to your learnON title at www.jacplus.com.au. *Note*: Question numbers may vary slightly.

Remember
1. What does FIFO mean?
2. What is the difference between a *tree changer* and a *sea changer*?

Explain
3. Refer to figure 1. Explain the difference between Australia's Heartland and its Frontier.
4. Look carefully at figure 1 and explain how the gap between Australia's east and west is predicted to alter over the next 40 years.

Discover
5. List the positive and negative factors of making a tree *change* or sea *change* as a:
   (a) family with young children
   (b) retired couple.
6. Convert table 1 into a pie graph, either using Excel software or by hand using a protractor. Describe the patterns that you can identify in your graph.

Predict
7. Do you think that the patterns you described in question 6 will still be evident in five years' time? Explain your answer.

Think
8. A more recent population migration is towards high-rise apartment living in the centre of major cities. How might this trend impact on these new residents and the sustainability of the environment their migration is creating? Use examples to justify your stance.

8.12 Why are people on the move in China?

8.12.1 Reasons for rural–urban migration

China has been experiencing a changing population distribution. The country's urban population became larger than that of rural areas for the first time in its history in 2012, as rural people moved to towns and cities to seek better living standards. China has become the world's largest urban nation.

Chinese labourers from the provinces have been moving to coastal cities in search of job opportunities, following reforms in 1978 which opened up China to foreign investment. Until then, rural–urban migration was strictly forbidden in China. Since then, more than 150 million peasants have migrated from the inner provinces to cities, mainly on the east coast. About half of rural migrants moved across provinces. This is the largest migration wave in human history (see figure 1).

FIGURE 1 People from Chinese inland provinces with lower wages and Human Development Index (HDI) values have moved to cities and provinces with higher HDIs and incomes.

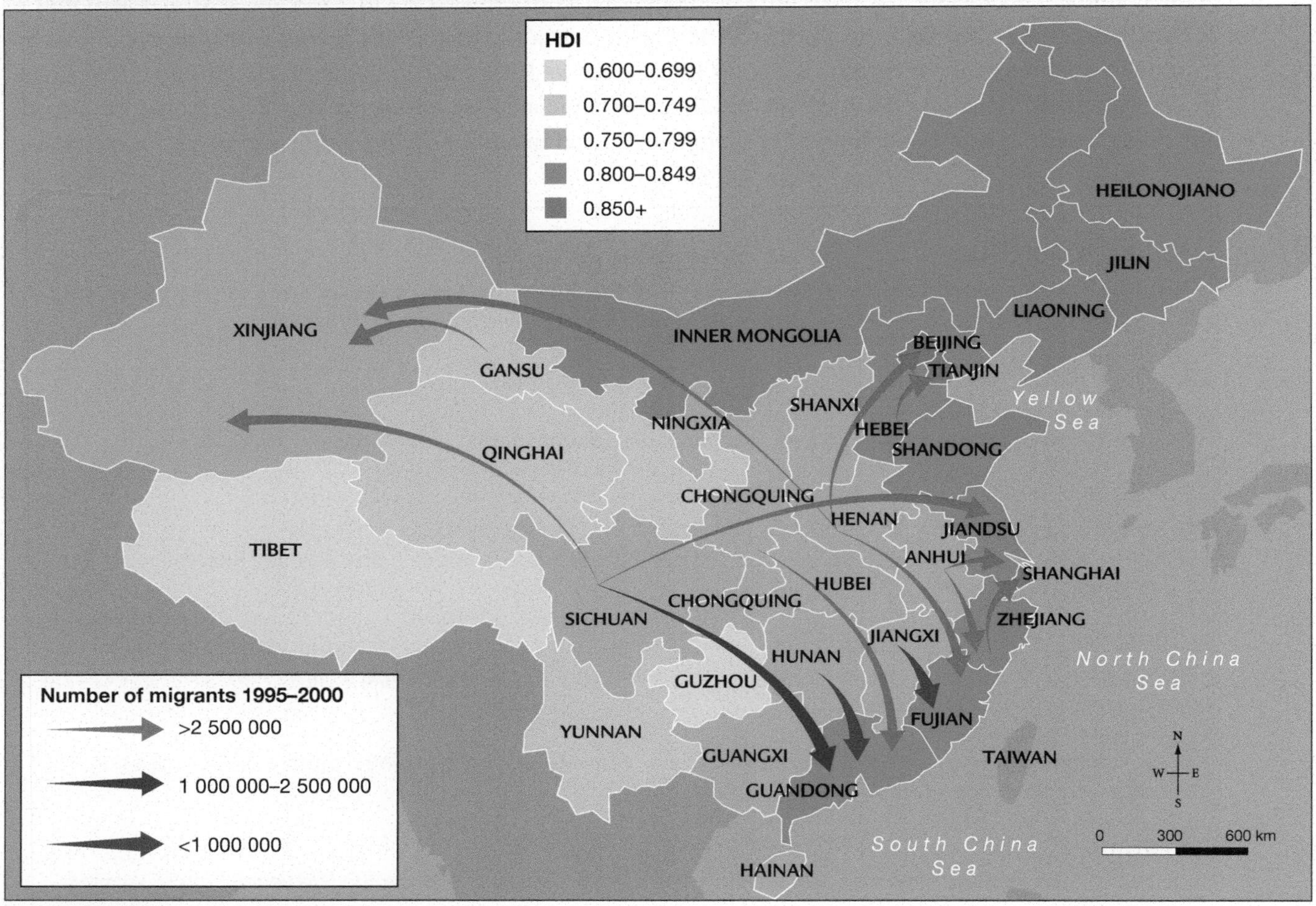

Source: Spatial Vision

Pull factors

Migrants from rural areas are attracted to urban regions largely for economic reasons — a higher income is achievable in a city (see figure 3). The average income of rural residents is about one-fifth that of urban residents on the east coast of China. Social factors are also important, with more opportunities for career development being available in cities; many people also desire a more modern urban lifestyle, with the benefits brought about by access to improved infrastructure and technology.

Push factors

Increasing agricultural productivity since the late 1970s has resulted in fewer labourers being needed on farms and thus a huge surplus of rural workers. These people have been forced to move to more urban areas in order to find employment. Agricultural production has meanwhile become less profitable, so workers have again been driven to cities to try to improve their economic situations (see figure 2).

FIGURE 2 A dramatic rural–urban migration shift has been occurring in China. In 2010, over half of China's population lived in rural areas, but by 2014 it was 46 per cent.

FIGURE 3 In 2015, Shanghai's population was estimated to be 23.74 million.

Political factors are also influential. China's central planners have encouraged local leaders in poor regions to encourage people to move to the cities. Their slogan was 'the migration of one person frees the entire household from poverty'.

8.12.2 Consequences of rural–urban migration

- China's urban population rose from around 170 million people in 1978 to 540 million in 2004, and then to nearly 800 million in 2015.
- In 1949, 89 per cent of people lived in rural areas; by 1979 this figure had dropped to 81 per cent. In 2014 this figure was 46 per cent.
- It is expected that, within 20 years, only 25 per cent of China's population will be living in rural areas, while the number of city-dwellers will rise to well over 1 billion people.
- Some people predict that by 2025, China will have 15 super-cities with an average population of 25 million people each.
- It is thought that the number of people living in China's countryside could shrink from 500 million to 400 million people.
- Labourers from rural regions working in cities have to leave their families for months at a time or more.
- Tens of millions of people are classified as rural dwellers, even though they spend most or all of their time working in the cities. These people are denied access to social services, including subsidised housing, income support and education for their children.
- A shift to an increased urban population results in reduced population pressures on the land.
- Up to 40 per cent of rural income comes from urban workers sending money to their families at home.

8.12 Activities

To answer questions online and to receive **immediate feedback** and **sample responses** for every question, go to your learnON title at www.jacplus.com.au. *Note*: Question numbers may vary slightly.

Remember

1. How has the percentage of people living in China's rural areas *changed* since 1949? What is this number expected to be in the future?
2. Describe the main *changes* that have occurred within China's urban population since 1978.

Explain

3. Explain in your own words the main reasons for the dramatic *change* in China's population distribution.
4. Classify each of the various consequences of this *change* as positive or negative.

Discover

5. Use the **China's urban growth** weblink in the Resources tab to respond to the following:
   (a) Describe population *changes* in the various cities in China.
   (b) 'The largest population growth has occurred in cities on China's coastline.' How true is this statement? Explain your answer using figures from the website.

Think

6. Creatively (in graphic or diagrammatic form) present some of the dramatic statistics in this subtopic to inform others of the *scale* of the *changes* happening to the distribution of China's population.

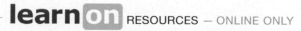 **RESOURCES** — ONLINE ONLY

- **Try out this interactivity:** Urban/rural China (int-3116)
- **Explore more with this weblink:** China's urban growth

 Deepen your understanding of this topic with related case studies and questions.
- China

8.13 Review

8.13.1 Review

The Review section contains a range of different questions and activities to help you revise and recall what you have learned, especially prior to a topic test.

8.13.2 Reflect

The Reflect section provides you with an opportunity to apply and extend your learning.

Access this subtopic at **www.jacplus.com.au**

TOPIC 9
The rise and rise of urban settlements

9.1 Overview

Numerous **videos** and **interactivities** are embedded just where you need them, at the point of learning, in your learnON title at www.jacplus.com.au. They will help you to learn the content and concepts covered in this topic.

9.1.1 Introduction

In 2008, for the first time in history, the majority of the world's population lived and worked in towns and cities. This urban population is projected to continue growing in the future. The fast pace and unplanned nature of this growth has seen the development of megacities — and along with opportunities come many problems. It is a challenge to create sustainable urban environments that meet the needs of the people living in these places.

Smog over the city of Shanghai

9.2 Where are the world's cities?

9.2.1 Where are cities located?

How is a city different from other urban areas such as towns and villages? A city is a large and permanent settlement, and is usually quite complex in terms of transport, land use and **utilities** such as water, power and **sanitation.**

The image of the Earth at night (figure 1) shows where lights are shining. The brightest areas on the map are the most urbanised, but might not be the most populated. If you compare this image with figure 3, you can make some comparisons. For example, there are very bright lights in western Europe (Belgium, The Netherlands, France, Spain and Portugal, Germany, Switzerland, Italy and Austria) and yet more people living in China and India. Refer to your atlas to locate these countries.

The world's cities are generally located along or close to coastlines and transport routes. Some regions remain thinly populated and unlit. Antarctica is entirely dark. The interior jungles of Africa and South America are mostly dark, but lights are beginning to appear there. Deserts in Africa, Arabia, Australia, Mongolia and the United States are poorly lit as well, although there are some lights along coastlines. Other dark areas include the forests of Canada and Russia, and the great mountains of the Himalayan region and Mongolia.

FIGURE 1 Satellite image of the Earth at night

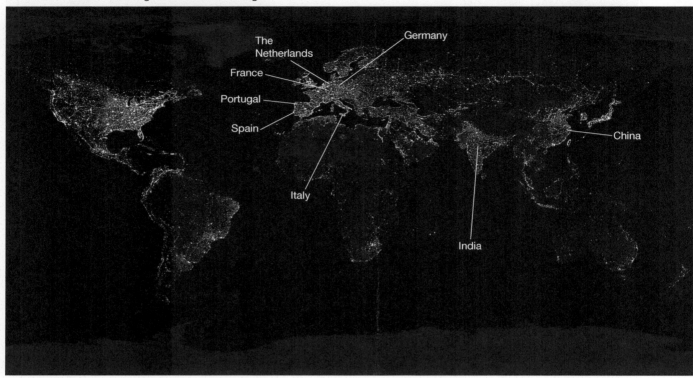

FIGURE 2 Medellin, the second-largest city in Colombia, South America

FIGURE 3 Distribution of urban centres, 2015, with selected city populations

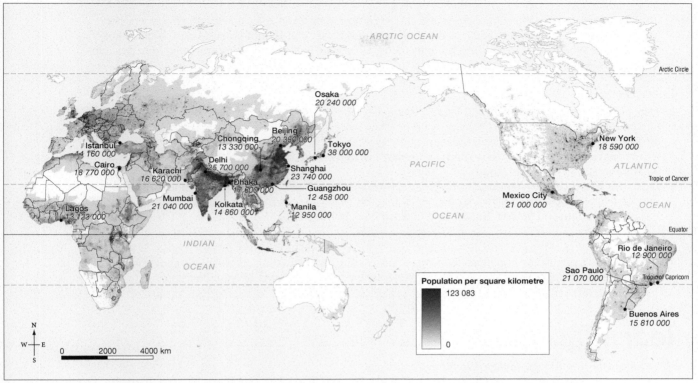

Source: United Nations. Department of Economic and Social Affairs. Population Division (2014). World Urbanization Prospects: The 2014 Revision. CD-ROM Edition.

9.2 Activities

To answer questions online and to receive **immediate feedback** and **sample responses** for every question, go to your learnON title at www.jacplus.com.au. *Note*: Question numbers may vary slightly.

Remember

1. How is a city different from a town or a village?
2. What do the bright lights in figure 1 show?

Explain

3. Study figures 1 and 3 and refer to a political map in your atlas. Which of the following statements are true and which are false? Rewrite the false statements to make them true.
   (a) Japan is a highly populated country with many cities.
   (b) The west coast of the United States is more densely populated than the east coast.
   (c) The Amazon rainforest does not have any settlements.
   (d) The eastern region of China has more cities than the western region.
   (e) The main city settlements in Australia are along the east coast.
   (f) The distribution of cities across Europe is uneven.
4. After completing the 'Describing photographs' SkillBuilder in subtopic 9.3, complete the following questions about figure 2.
   (a) Describe the foreground and background shown in the photograph.
   (b) List the natural and human characteristics shown in the photograph.
   (c) What does this photograph show about urban *environments*? How has the urban *environment* changed the natural *environment*?
   (d) How might the changes described in part (b) lead to an increased risk of erosion? (See topic 2 for information on erosion processes.)

(e) Imagine that the population of this city continues to increase. Describe what might happen to the land in the future.

(f) Do you think that all land surrounding cities should be able to be taken up by buildings? Why or why not?

(g) Investigate the place where you live. Are there land-use zones that cannot be built upon, such as 'green wedges'? Where are they and why are they there? Do you think they should be protected from development? Justify your answer.

Discover

5. Use a political map in your atlas and figure 1 to identify the following.
   (a) The Nile River
   (b) The Trans-Siberian railway from Moscow to Vladivostok
   (c) Highways linking cities in the western and eastern United States
   (d) The Himalayan mountain range
6. Go to the **World City Populations** weblink in the Resources tab. Work as a team of five and investigate the change in city population in different continents. Discuss which regions each member will investigate and record maps, data and graphs. Report your findings back to the group.

Predict

7. 'As the world's population continues to increase, cities will spread into the darker regions shown in figure 1.' State whether you agree or disagree with this statement, providing reasons for your decision.

 RESOURCES — ONLINE ONLY

🔗 **Explore more with this weblink:** World City Populations

9.3 SkillBuilder: Describing photographs

WHAT IS MEANT BY 'DESCRIBING A PHOTOGRAPH'?

A description is a brief comment (up to a paragraph) on a photograph, identifying and communicating features from a geographic point of view. As geographers, we use our understanding of the world to interpret the image and tell others about the main features or information the photograph reveals.

FIGURE 1 A modern city environment

Go online to access:
- a clear step-by-step explanation to help you master the skill
- a model of what you are aiming for
- a checklist of key aspects of the skill
- a series of questions to help you apply the skill and to check your understanding.

---

learn on RESOURCES — ONLINE ONLY

🎞 **Watch this eLesson:** Describing photographs (eles-1660)

🧩 **Try out this interactivity:** Describing photographs (int-3156)

9.4 SkillBuilder: Creating and reading compound bar graphs

WHAT ARE COMPOUND BAR GRAPHS?

A compound bar graph is a bar or series of bars divided into sections to provide detail of a total figure. These bars can be drawn vertically or horizontally. Compound bar graphs allow us to see at a glance the various components that make up the total.

Go online to access:

- a clear step-by-step explanation to help you master the skill
- a model of what you are aiming for
- a checklist of key aspects of the skill
- a series of questions to help you apply the skill and to check your understanding.

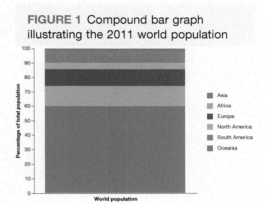

FIGURE 1 Compound bar graph illustrating the 2011 world population

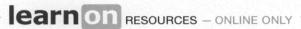

learn on RESOURCES — ONLINE ONLY

📄 **Watch this eLesson:** Creating and reading compound bar graphs (eles-1705)

🧩 **Try out this interactivity:** Creating and reading compound bar graphs (int-3285)

9.5 Why do people move to urban areas?

9.5.1 Factors causing people to move

There are many and varied reasons for people migrating to urban locations. These reasons are usually a combination of push and pull factors. Some people are 'pushed' from rural to urban areas within their own country. Others will travel from other countries to urban areas, 'pulled' by better opportunities.

Push factors

Geographical inequality is mostly responsible for the **migration** of people from rural to urban areas. **Push factors** that drive people towards cities usually involve a decline in living conditions in the rural area in which the people live. There are various situations that can cause this, including a decrease in the quality of agricultural land (caused by factors such as prolonged drought, erosion or desertification); poverty; lack of medical services or educational opportunities; war; famine from lack of food and/or crop failure; and natural disasters.

Pull factors

Pull factors refer to the attractions of urban areas that make people want to move there. Urbanisation in any country generally begins when enough businesses are established in the cities to provide many new jobs. Pull factors include job opportunities; better housing and infrastructure; political or religious freedom; improved education and healthcare; activities and enjoyment of public facilities; and family links.

FIGURE 1 Examples of push factors include lack of medical services, war, crop failure, prolonged drought and desertification, famine, poverty and lack of educational opportunities.

FIGURE 2 Examples of pull factors include religious tolerance, improved healthcare, job opportunities, family links, better housing and infrastructure, political freedom and better educational opportunities.

9.5.2 India

Rural–urban migration in India

More than 250 million Indian people migrate within their country at some time in their life. Although most movements are only short distances from the family home, more than 40 million Indians move to another state. The largest number of people (around 40 per cent) are attracted to the largest cities — Delhi and Mumbai. These cities are located in the richer states (those with a higher Gross State Domestic Product, or GSDP), therefore attracting more workers. The population density of Mumbai in 2015 was 20 694 people per square kilometre, while that of Delhi was 25 535.

FIGURE 3 Urban and rural migration in India

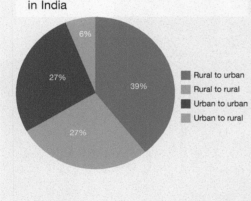

- ■ Rural to urban
- ■ Rural to rural
- ■ Urban to urban
- ■ Urban to rural

FIGURE 4 Migration flows in India

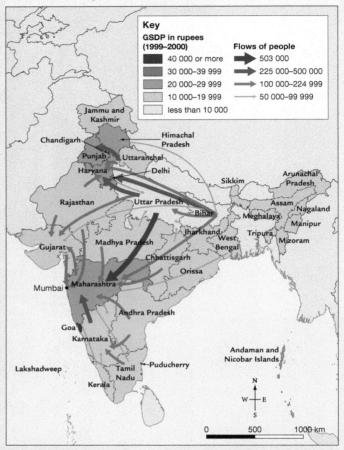

Source: The Atlas of Human Migration by Russell King et al © Myriad Editions | www.myriadeditions.com. p. 54 of The Atlas of Human Migration UK Edition

9.5.3 Which cities attract workers?

Taxi drivers, construction workers, teachers, nurses, house cleaners, accountants, nannies — there are many job opportunities for both skilled and unskilled workers that attract people to cities. These people may come from a different area within a country or across borders from different countries.

'Gateway cities' are arrival points for many migrant workers. These cities are large enough to provide many different jobs and are therefore attractive to people moving from other regions. Some cities, such as Dubai, are reliant on their foreign workers.

More than two-thirds of Dubai's population is migrant labour, with many working in building construction. These labourers — mostly from India, Pakistan and Bangladesh — live in migrant camps that can be up to two hours away from the work site. They are often poorly paid.

TABLE 1 The 25 gateway cities with the largest foreign-born population

Continent	Gateway city/cities
North America	Toronto, Vancouver, New York, Los Angeles, Chicago, Boston, Washington DC, Miami, Dallas, Houston, San Francisco, Riverside, New Jersey
South America	Buenos Aires
Europe	London, Paris, St Petersburg, Moscow
Africa/Middle East	Jeddah, Riyadh, Dubai, Abu Dhabi, Tel Aviv-Yafo
Asia	Hong Kong, Singapore
Oceania	Sydney, Melbourne, Perth, Auckland

9.5.4 Africa

CASE STUDY

Growth of cities in Africa

Africa now has a larger urban population than North America and has 25 of the world's fastest-growing large cities — the number of people living in cities in Africa is increasing by about one million every week. Some of Africa's cities are expected to grow by 85 per cent by 2025. By 2050, the urban population is expected to triple from 400 million people to 1.2 billion. Over half of the urban population is below the poverty line in Angola, Chad, Madagascar, Malawi, Mozambique, Niger, Sierra Leone and Zambia. In many other countries, including Burundi, Gambia, Kenya and Zimbabwe, 40–50 per cent of the population are living below the poverty line.

In most African cities, between 40 and 70 per cent of the population live in slums or squatter settlements. In cities such as Nairobi, Lagos, Cairo and Rwanda, 60–70 per cent of the population live in slum conditions, which occupy about five per cent of the land.

FIGURE 6 The projected growth of African cities from 2010 to 2025

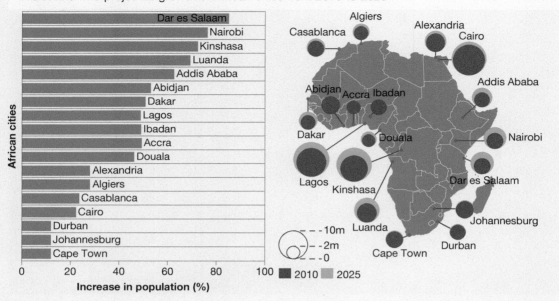

FIGURE 7 A slum in Nairobi

9.5.5 Regional differences

During the nineteenth and early twentieth centuries, urbanisation occurred because of migration and the growth of industries. New job opportunities in the cities attracted people from rural areas and migrants provided a cheap workforce for factories. At that time, death rates in cities were high because they were unhealthy places (with crowded living conditions, diseases and a lack of sanitation) and urban growth was slow. Workers often found it hard to find somewhere to live — it was not unusual for an entire family to be living in a single room. In many European cities the number of deaths was higher than the number of births, and migrants provided most of the population growth. London was a typical example of this, as was New York, where urban growth has slowed.

It is a very different experience in developing countries. Most urban growth results from natural increase; that is, people being born in cities, rather than migrating to cities. With the additional population increase caused by migration from rural areas in search of better jobs, many cities in Asia and Africa have exploded in size.

Cities can be great places and should not be viewed negatively. For example, people can more easily access basic services in urban areas than in rural areas so, although poverty may be present in urban environments, cities also offer an escape from poverty. Cultural activities are often enhanced in cities that attract migrants from many different areas — food and music are obvious examples. There also tends to be a greater tolerance of different migrant and racial groups living close together.

9.5 Activities

To answer questions online and to receive **immediate feedback** and **sample responses** for every question, go to your learnON title at www.jacplus.com.au. *Note*: Question numbers may vary slightly.

Remember

1. What are push factors? What are pull factors? Give two examples of each.
2. Match each of the images in figures 1 and 2 with the push or pull factors listed in the captions.
3. What is a gateway city? Why are people attracted to them?
4. Watch the video about urbanisation in **Mongolia** in *myWorld Atlas*.
   (a) List the push and pull factors that have caused people to move to Ulaanbaatar.
   (b) Describe the living conditions of these people. Do you think they are better or worse than their living conditions in rural areas? Justify your response.

Explain

5. What does figure 3 tell you about the urban and rural movement of Indian people?
6. Figure 4 shows a relationship between the movement of people and the wealth (indicated by GDP) of the states they are moving to and from.
   (a) Have people moved to or from low-GDP states? Give two examples.
   (b) Where are the main flows of people occurring? How many people moved to Maharashtra?
   (c) What might be the main reason for movements from poor states to rich states?
7. Study figure 6 and refer to an atlas map of Africa.
   (a) Name the three largest African cities in 2010 and the three predicted to be largest in 2025. In which countries are they located?
   (b) Describe the distribution of Africa's large cities. How many are inland? How many are on the coast? Which are located in the north, south-east and west of the continent? List the countries that do not have large cities.
   (c) What does it mean to live below the poverty line? Locate the cities in which more than half the population is living below the poverty line.
8. What is the difference between urban population increase from migration and from natural increase? Which of these is more likely to occur in a city located in a developing country? Why?

Discover

9. Use an atlas to locate all the gateway cities mentioned in table 1, and then mark their locations on a blank map of the world. When your map is complete, describe the distribution of the major gateway cities around the world.
10. Find out the population density of the capital city in your state or territory. How does it compare to that of Mumbai and New Delhi? List all the ways in which living in one of these Indian cities might be different to life in your local city.

▶

Think

11. Look at figure 7. Draw a sketch of this scene and annotate it with geographical questions you would like answered about the *environment* and the people living there.
12. What do you think is the future *sustainability* of the *place* shown in figure 7, especially if the population of this city is going to increase?

 RESOURCES — ONLINE ONLY

♦ **Try out this interactivity:** Urban push and pull factors (int-3118)

 myWorldAtlas Deepen your understanding of this topic with related case studies and questions.
   ⊙ Mongolia

9.6 How do urban areas affect people's ways of life?

9.6.1 The wealth of cities

Both small and large urban areas can provide people with positive and negative experiences.

Cities attract people to them with the opportunity of work and the possibility of better housing, education and health services. There is a strong interconnection between the wealth of a country and how urbanised it is. Generally, countries with a high **per capita income** tend to be more urbanised, while low-income countries are the least urbanised.

This happens because people grouped together create many chances to move out of poverty, generally because of increased work opportunities. There are often better support networks from governments and local councils. It is also cheaper to provide facilities such as housing, roads, public transport, hospitals and schools to a population concentrated into a smaller area.

9.6.2 Urban challenges

Rapid population growth in urban areas can result in problems such as poverty, unemployment, inadequate shelter, poor sanitation, dirty or depleted water supplies, air pollution, road congestion and overcrowded public transport.

Slums

In many developing countries, urban growth has resulted in unplanned settlements called **slums** (other terms used around the world include ghettos, favelas, shantytowns, bidonvilles and bustees). Almost 1 billion people live in slums worldwide.

The United Nations defines a slum as follows.

> … one or a group of individuals living under the same roof in an urban area, lacking one or more of the following five amenities: (1) durable housing (a permanent structure providing protection from extreme climatic conditions); (2) sufficient living area (no more than three people sharing a room); (3) access to improved water (water that is sufficient, affordable and can be obtained without extreme effort); (4) access to improved sanitation facilities (a private toilet, or a public one shared with a reasonable number of people); and (5) secure tenure and protection against forced eviction.

Water and sanitation

Many cities cannot keep up with more and more people living in urban areas, which means it is difficult to provide water and toilets for everyone (see table 1). Without these services, more people suffer from diseases and poor health and are unable to go to work or school.

TABLE 1 Availability of water and sanitation in selected regional cities around the world

	Water on premises (%)		Flush toilets (%)	
	Urban poor	Urban non-poor	Urban poor	Urban non-poor
Latin America	59	74	44	67
Sub-Saharan Africa	31	46	20	32
South, Central and West Asia	59	74	48	60
South-east Asia	36	50	67	88

Transport and pollution

In cities that can't keep up with rapid population growth, traffic congestion and overcrowded public transport mean that many people must travel for hours to get to and from work (see figure 1).

Pollution is also a problem that affects the health of people living in cities. Most cities have high levels of air pollution and some — including Mexico City, Buenos Aires, Beijing and Los Angeles — are famous for being so polluted.

According to the World Bank, 16 of the world's 20 cities with the worst air are in China. The burning of coal is the main source of air pollution in China.

FIGURE 1 Traffic congestion in Los Angeles, United States

FIGURE 2 The proportion of each country's urban population living in slums

The proportion of each country's urban population living in slums

0–10%		50–60%	
10–20%		60–70%	
20–30%		70–80%	
30–40%		80–90%	
40–50%		90–100%	
No data			

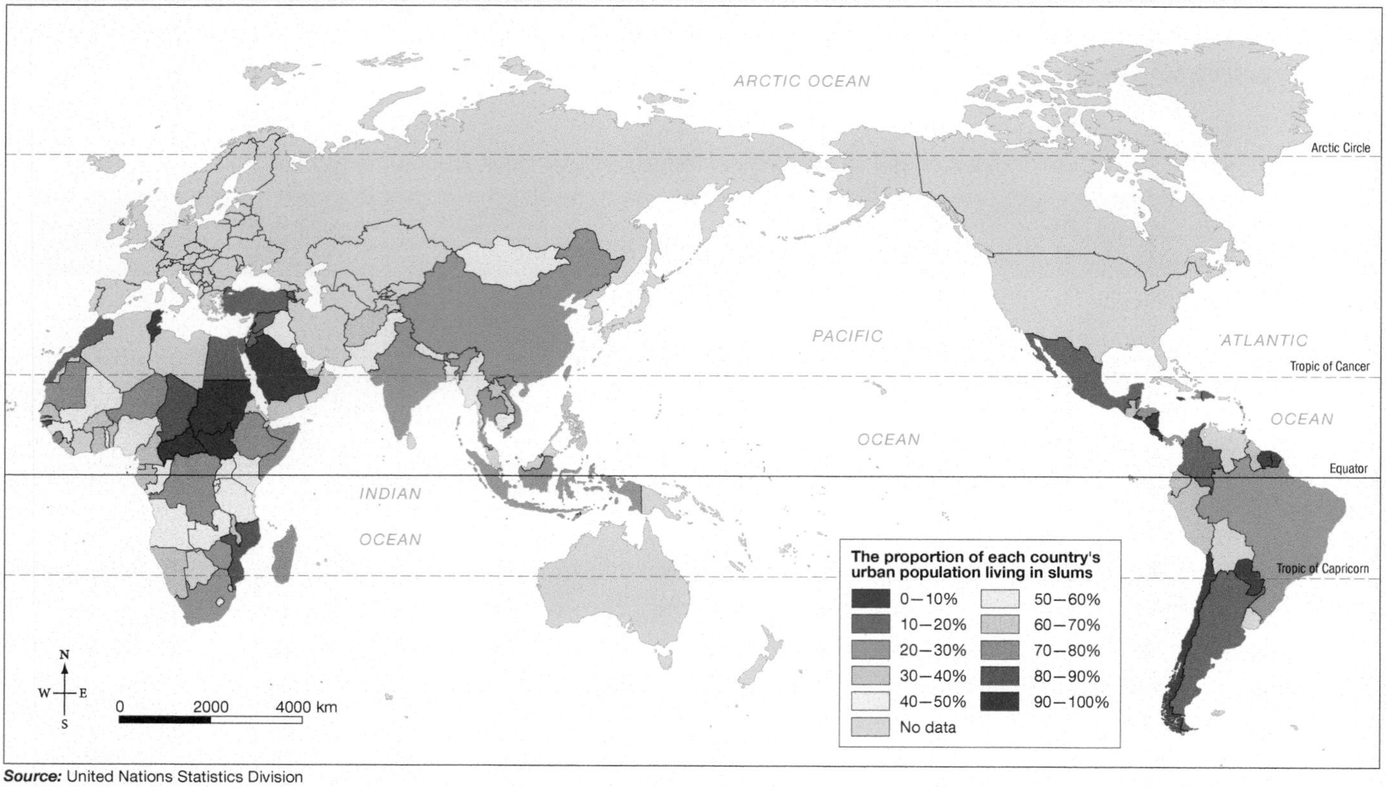

Source: United Nations Statistics Division

9.6 Activities

To answer questions online and to receive **immediate feedback** and **sample responses** for every question, go to your learnON title at www.jacplus.com.au. *Note:* Question numbers may vary slightly.

Remember

1. What is a slum? Make a list of some other names for slums.
2. Why are transport and pollution often problematic in large urban areas?

Explain

3. Why is it difficult in a country the size of Australia, with population concentrated on the coast, to provide services in outback areas? How would providing services be different in a country such as Luxembourg in Europe? Look at the size of Luxembourg in an atlas or by using Google Maps or Google Earth.
4. Study figure 2.
   (a) In which continent are the most urban slums found?
   (b) Name three countries in this continent with very high numbers of slums.
   (c) Describe the general pattern shown in the map.
5. Draw a graph to illustrate the water and sanitation data for urban areas given in table 1. Describe the relationship between poor and non-poor people in urban areas and their access to clean water and toilets.

Discover

6. Use figure 3 in subtopic 9.2 and the *myWorld Atlas* statistical mapper to find the relationship between urbanisation and wealth. Give five examples of countries from different continents that are highly urbanised and wealthy, and five that are not urbanised and poor. Do any countries not fit this pattern? Name them.

7. Which Australian city do you think would have the worst:
   (a) transport problems
   (b) pollution?
8. Justify your choice. Conduct some research to find out which city has the worst data for these two urban problems.

Predict

9. Imagine you live in a poor rural village in India with no education or work. List the possible attractions of moving to an urban area.

 **Deepen your understanding of this topic with related case studies and questions.**
   ◐ Mexico City

9.7 SkillBuilder: Constructing a basic sketch map

WHAT IS A BASIC SKETCH MAP?

A basic sketch map is a map drawn from an aerial photograph or developed during field work that identifies the main features of an area. Basic sketch maps are used to show the key elements of an area, so other more detailed characteristics are not shown.

Go online to access:

- a clear step-by-step explanation to help you master the skill
- a model of what you are aiming for
- a checklist of key aspects of the skill
- a series of questions to help you apply the skill and to check your understanding.

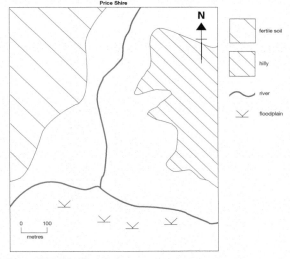

FIGURE 1 Sketch map of land use in Price Shire

Source: Price Shire Council

 ONLINE ONLY

    Watch this eLesson: Constructing a basic sketch map (eles-1661)

    Try out this interactivity: Constructing a basic sketch map (int-3157)

9.8 Where are the world's megacities located?

9.8.1 What is a megacity?

Over the next century, urbanisation is predicted to increase at an even greater rate than it has in the past. Forty years ago, only one-third of the world's population lived in urban areas — now this figure is just over one half. Many of these people are attracted to cities with huge populations, and increasingly these cities are becoming megacities.

A **megacity** is a city with more than 10 million inhabitants. When you consider that Australia's population is around 23 million — with around 4.5 million of those living in our largest city, Sydney — it is hard to imagine what it would be like to live in a megacity.

The number of megacities has grown over time. In 1950, only two cities in the world — Tokyo and New York — had a population above 10 million. By 1975 there were four; by 2000 there were 17, and in 2016 there were 36 megacities. By 2030, it is predicted that there will be over 41 megacities in the world. Nineteen of these cities exceed the megacity definition and have a population greater than 15 million.

The distribution of megacities — that is, where they are located over space in the world — has also changed. In 1975, two megacities were located in the Americas and two in Asia. In 2014 more than half (15) of all megacities were located in Asia; and it is predicted that, in 2030, 23 of the 41 megacities will be located in Asia. There is also a change in terms of the wealth of countries that contain megacities, with the majority now located in developing countries. This is in contrast to the development of urbanisation, in which North America and Europe were the focus of historic urban growth. By 2030, it is predicted that 23 megacities will exist in less developed countries.

FIGURE 1 The growth of megacities over time

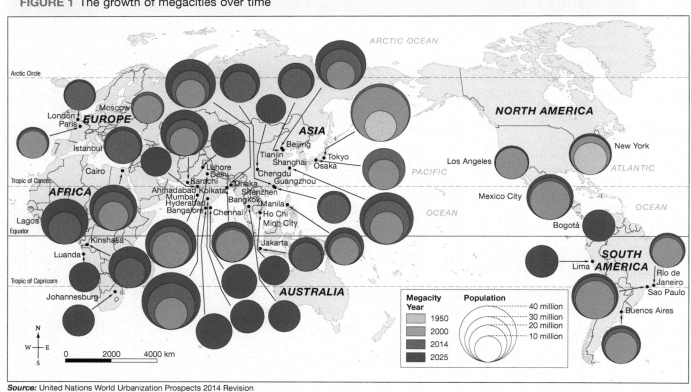

Source: United Nations World Urbanization Prospects 2014 Revision

9.8.2 The never-ending city

In some parts of the world, megacities are merging to create **megaregions**. These regions are home to huge populations. Examples of megaregions include:

- Hong Kong–Shenzhen–Guangzhou in China, already home to around 120 million people
- Kyoto–Osaka–Kobe, with a population of over 20 million in 2010
- Rio de Janeiro–São Paulo in Brazil, with a population of more than 23 million people.

Hong Kong–Shenzhen–Guangzhou

This region is located in southern China on the South China Sea. The three main cities are Hong Kong, Shenzhen and Guangzhou, which are linked by transport routes and provide great economic opportunities. Until 1979, Shenzhen was a fishing village. In 1980 the government declared the area to be a special economic zone, attracting businesses and investment from other countries. Since then, the area has undergone rapid urbanisation that has dramatically changed the landscape around the Pearl River delta (see figure 2).

FIGURE 2 Change in the Pearl River delta between (a) 1988 and (b) 2014

Source: NASA

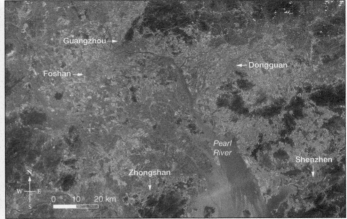

Source: NASA

FIGURE 3 The city of Shenzhen, in the Pearl River delta, in the twenty-first century

In 1988, the rivers and streams flowed through a fertile region with rice paddies, wheat fields, orchards and fish ponds. The region was mostly rural, and the population of roughly 10 million distributed between rural areas and a few cities.

By 2014 these cities have grown quickly and merged into an interconnected megalopolis with a population of 42 million. When combining the population of Hong Kong, the total is around 120 million.

Megacity facts

- Over half the future growth in megacities will be within Asia.
- The 20 largest cities consume 80 per cent of the world's energy and produce 80 per cent of global greenhouse gas emissions.
- Slums in megacities are especially vulnerable to climate change, as they are often built on hazardous sites in high-risk locations.

9.8 Activities

To answer questions online and to receive **immediate feedback** and **sample responses** for every question, go to your learnON title at www.jacplus.com.au. *Note*: Question numbers may vary slightly.

Remember

1. What is a megacity? How many megacities were there in 2010?
2. Name the first two megacities and the countries where they are located.
3. What is a megaregion?

Explain

4. Refer to figure 1. Describe how the number and distribution of megacities has changed over time.

Discover

5. Use an atlas to locate the three megaregions mentioned in section 9.8.2. Why do these regions develop?
6. Study figure 2. Research the 'dead zone' in the sea at the mouth of the Pearl River. What does this mean, and what is its cause?

Predict

7. Describe the changes that have occurred in the Hong Kong–Shenzhen–Guangzhou region. Find this *place* in an atlas and describe where it is in relation to the rest of China and to two other countries in Asia.
8. What impact will this urban area have on people and the *environment*?

Think

9. Work with another student to produce a Prezi or PowerPoint presentation or an animation showing the world's megacities in 2014 and 2030. Include images from the internet and data from figure 1. You may like to choose appropriate music to accompany the presentation.

 ONLINE ONLY

 Try out this interactivity: Megacity march (int-3119)

9.9 What are the causes and effects of Indonesia's urban explosion?

9.9.1 Introduction

Many people do not realise that the fourth most populated country in the world is one of our nearest neighbours. Like many countries in Asia, Indonesia has experienced rapid urban growth, but this has occurred only relatively recently.

Indonesia's population of more than 257 million people (2015) lives on a chain or cluster (an archipelago) of more than 18 000 islands (see figure 1). However, its population is not evenly distributed. Only about 11 000 of the islands are actually inhabited. Sixty per cent of Indonesia's population is concentrated on only seven per cent of the total land area — on the island of Java.

FIGURE 1 Map of Indonesia

Source: Spatial Vision

Indonesia has changed from a rural to an urban society quite recently. In 1950, only 15.5 per cent of its population lived in urban areas. In 2014, this had increased to 53 per cent.

Like many countries in Asia, Indonesia has a high concentration of its urban population in a few large cities. In 1950, there was only one city that was home to more than one million people in Indonesia: Jakarta. That had increased to four cities by 1980, eight by 1990, 10 by 2000 and 11 by 2015. In fact, more than one-fifth of the Indonesian urban population now lives in the Jakarta metropolitan area (JMA).

9.9.2 Causes of urbanisation

More than one-third of Indonesia's urban population growth resulted from natural increase. It took until 1962 for Indonesia's population to reach 100 million people. However, it then took only until 1997 to reach 200 million. In the early 1970s, Indonesia's birth rate was very high — 5.6 children per woman. Although the growth rate has fallen dramatically from 2.3 per cent in 1970 to about 1.2 per cent in 2012. Between 2005 and 2010 about 4 464 000 were born in Indonesia each year.

The Indonesian government placed few restrictions on rural–urban migration. Most of the migration movement consisted of the rural poor moving into cities and especially into slums, leaving their families behind in the villages. On top of this, in recent years about 20 000 foreigners per year have obtained work permits for Indonesia.

FIGURE 2 Selected south-east Asian city populations, 1950–2030 projected

Source: United Nations, Department of Economic and Social Affairs, Population Division (2014). World Urbanization Prospects: The 2014 Revision, CD-ROM Edition

Investment from within Indonesia and from other countries has tended to occur mainly in the large urban areas, because these areas can supply the workers, transport (by sea and over land), water and electricity that are needed by industry.

The first president of Indonesia wanted Jakarta to be like the world's great cities, such as Paris and New York, as well as a focus for other Indonesian cities. President Sukarno therefore built broad avenues, highways and electric railway lines, luxurious housing estates, high-rise buildings, universities and industrial estates in Jakarta.

FIGURE 3 The Jakarta metropolitan area has a population density of 14 464 people per square kilometre, and is the second largest urban area in the world.

9.9.3 Consequences of urbanisation

Growth of Jakarta

One of the consequences of urbanisation in Indonesia has been the dramatic growth of Jakarta, Indonesia's capital and largest city, located on the north-west coast of Java. The central island of Java is the world's most populous island, having a population density of 1000 people per square kilometre. The JMA is now one of the world's largest urban areas. In 1930, Jakarta's population was around half a million people. By 1961 it had grown almost six-fold to 2.97 million. By 2005, it was almost 9 million. Today, the total population of the Jakarta urban area is more than 31 million and the city controls around 70 per cent of Indonesia's economy.

FIGURE 4 Jakarta's urban growth

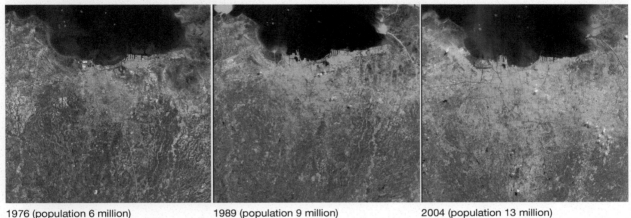

1976 (population 6 million)   1989 (population 9 million)   2004 (population 13 million)

FIGURE 5 Smog over Jakarta

Loss of land

Prime agricultural areas have been lost and become residential and industrial areas. Urban land is worth more than agricultural land.

As Jakarta has become more urbanised, there has been a decrease in the amount of open green space — from nearly 30 per cent of the city's total area in 1984 to less than 10 per cent today.

Environment

Indonesia's level of sewerage and sanitation coverage is very low. Sewage from houses and from industry as well as industrial effluents and agricultural run-off are polluting surface and groundwater. Air pollution levels are high, with traffic and industrial fumes combining with smoke from fires set by farmers and plantation owners in rural areas clearing forest lands for agricultural use.

FIGURE 6 Traffic congestion in Jakarta

Food production

Because young people, especially young men, migrate to Indonesia's cities in search of better job opportunities, there are fewer people taking over their families' farms. This could lead to the possibility of a food crisis if food production levels are not increased.

Job opportunities

Labourers who lived in Java and did not own land used to have very few sources of income. Now, most landless rural families on Java have at least one person working outside the village in a factory or service job. Today, less than 20 per cent of households depend on agriculture for their livelihood.

Subsidence

Land has been subsiding because more groundwater is being extracted, and also because of the additional load that the ground has to bear because of an increased volume of construction. Subsidence causes cracking of buildings and roads, changes in the flow of rivers, canals and drains, and increased inland and coastal flooding. In some parts of Jakarta, land has subsided by 1–15 centimetres per year — in other areas, this has been up to 28 centimetres per year.

New urban areas

New towns and large-scale residential areas have been developed in and around Jakarta. However, heavy flows of commuter traffic have led to increased levels of traffic congestion between the scattered new towns and the cities.

9.9 Activities

To answer questions online and to receive **immediate feedback** and **sample responses** for every question, go to your learnON title at www.jacplus.com.au. *Note*: Question numbers may vary slightly.

Remember

1. (a) What is Indonesia's current population? If the area of Indonesia is 1 904 569 square kilometres, what is its approximate population density?
   How does this compare to Australia's population density of 2.9 people per square kilometre?
   (b) Describe, using statistics, how Indonesia has become very urbanised in a relatively short time.

Explain

2. Explain, in your own words, why Indonesia has become very urbanised.
3. Explain how and why Jakarta has become a major city within Indonesia and also on a world *scale*.
4. Why do you think people have moved from rural areas to urban areas within Indonesia?
5. What is the *interconnection* between the increasing population in Indonesia and the subsidence of land?

Think

6. What do you believe are the three main reasons that Indonesia has undergone such rapid urbanisation? Give reasons for your choices.
7. Which of the consequences of urbanisation do you think may continue to have the biggest effects on the *environment* in the future? Why? How important are these considerations to you?
8. How is the urbanisation of Indonesia similar to and different from the urbanisation of another country you have studied, such as Australia, China or the United States?

 myWorldAtlas   Deepen your understanding of this topic with related case studies and questions.
❂ **Urbanisation in Indonesia**

9.10 What are the characteristics of cities in South America?

online only

Access this subtopic at **www.jacplus.com.au**

9.11 What are the characteristics of cities in the United States?

9.11.1 Cities in the United States

The distribution of major cities across the United States, including the 10 largest cities (by population), is shown in figure 1. The largest is New York City, New York, which is home to nearly 9 million people. The second-largest city is Los Angeles, California, with a population of almost 4 million; and the third-largest is Chicago, Illinois, with nearly 2.7 million people.

FIGURE 1 The distribution of major cities in the United States

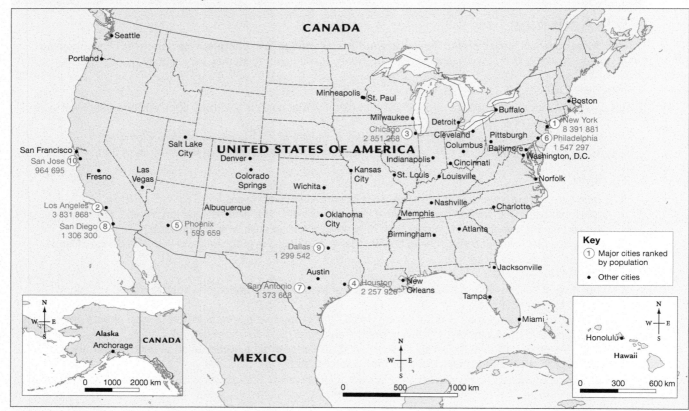

Source: Spatial Vision

9.11.2 New York City

The United States has a number of major cities distributed across the country. The largest of these is New York, one of the world's megacities as its metropolitan area includes New York–New Jersey–White Plains. Its population in 2014 was 18 591 million.

In 1950 there were only two megacities, and New York was one of them (Tokyo in Japan was the other). In 2014, New York was the ninth-largest city in the world. By 2030 it

FIGURE 2 Manhattan Island and Central Park in New York

is expected to be the fourteenth. There are only 11 states in the United States that are home to more people than New York City.

New York City is located on the eastern Atlantic Ocean at the mouth of the Hudson River. It is made up of five counties, or boroughs, separated by waterways — these are the Bronx, Brooklyn, Manhattan, Queens and Staten Island. Being located on four islands makes land very scarce and population density very high, at 10 194 people per square kilometre.

People

For many years, almost all immigrants came to the United States through New York City — and many of them remained. Many people living in New York are originally from European countries, but there are large numbers from the West Indies, South and Central America, the Middle East and eastern Asia. Around 800 languages are spoken in New York — around 36 per cent of the city's population were born overseas.

TABLE 1 Population statistics of New York City

City/Borough	2014 population (estimate)	Population density (people/km²)
City of New York	8 491 079	10 756
The Bronx	1 438 159	13 221
Brooklyn	2 621 793	14 182
Manhattan	1 636 268	27 673
Queens	2 321 580	8 237
Staten Island	473 279	3 151
State of New York	19 746 227	159

Economy

New York City is a major world centre of trade, commerce and banking (New York is also home to the largest stock exchange in the world), manufacturing, transportation, finance, communications, and culture and theatrical production. It is also the headquarters of the United Nations and a leading seaport.

Boroughs

The Bronx County is the only part of New York that is connected to the US mainland. Historically there were many Irish and Italian migrants; today they are mostly Russian and Hispanic.

Brooklyn (also known as Kings County) is where most New Yorkers live; but Manhattan is the most densely populated county. It contains the highest number of skyscrapers, and includes Central Park and the village of Harlem. Central Park is nearly twice as large as the world's second-smallest country, Monaco.

FIGURE 3 Geographical characteristics of New York City

Source: Created from data from City of New York, New Jersey Department of Environmental Protection, New Jersey Geographic Information Network 2012

9.11 Activities

To answer questions online and to receive **immediate feedback** and **sample responses** for every question, go to your learnON title at www.jacplus.com.au. *Note:* Question numbers may vary slightly.

Remember

1. Name the five boroughs or counties that make up New York.
2. In what year was New York one of the world's only two megacities?

Explain

3. Describe the distribution of major cities in the United States. Where are most located?
4. Describe New York's location within the United States and in terms of its natural geographical features. How have these features helped make New York a major city?

Discover

5. Conduct some research to find images of New York that reflect its characteristics. Use the information in this subtopic about its people and economy as well as the information in figure 3. Include images of buildings, transport, culture and businesses, and produce a collage with labels. This might be in an electronic format or produced as a poster.
6. Find the names of one newspaper, magazine, fashion brand, major bank, art gallery and theatre that are located in New York. Compare your findings with those of other students. Are these well-known businesses? How does this help make New York an important business and cultural centre?

Predict

7. Draw a sketch of figure 2. Use the map to help you label Central Park and the Hudson River. In which direction is the photographer facing?

Think

8. Use the data in table 1 to draw a bar graph showing the population and densities of New York and it's boroughs. Describe the pattern that you see.
9. Use figures 2 and 3 and the graph you created in question 8 to write two sentences about population density in New York and where growth might occur in the future.

 Deepen your understanding of this topic with related case studies and questions.
❯ **Urbanisation in Australia and the USA**

9.12 What are the characteristics of cities in Europe?

Access this subtopic at **www.jacplus.com.au**

9.13 How can cities become sustainable?

9.13.1 Sustainable urban solutions

Cities are huge consumers of goods and services. To be sustainable, cities need to develop so that they meet present needs and leave sufficient resources for future generations to meet their needs.

A sustainable city, or eco-city, is a city designed to reduce its environmental impact by minimising energy use, water use and waste production (including heat), and reducing air and water pollution.

Every city in the world experiences some type of problem that needs to be overcome — inadequate housing, urban sprawl, air and/or water pollution and waste disposal are just a few. Solutions to city problems have a better chance of succeeding if:

- responsibility is shared between governments, communities and citizens
- communities are involved in projects and decision making.

9.13.2 Sustainable urban projects

Urban greening program, Sri Lanka

Producing food in cities provides people with an income and improves local environments, as well as reducing the distance that food must travel to a consumer — '**food miles**'. With support from the Department of Agriculture, the Department of Education and the Youth Services Council, three city councils in Sri Lanka developed a program of community environmental management that led to the creation of 300 home gardens and 100 home-composting programs. It also helped organise and empower community groups, and the idea has now spread to many other municipalities in the country.

FIGURE 1 The urban greening program in Sri Lanka has been a success in many communities.

Beekeeping in urban areas

A worldwide movement of urban beekeeping has had beekeepers in partnership with businesses and property owners in major cities to place beehives on rooftops. The movement makes a strong connection between urban areas and food supply. This is happening in cities such as London (the Lancaster Hotel in London has its own hives, as does Buckingham Palace), New York, San Francisco, Paris, Berlin and Toronto. In Australia, there is a growing number of hives on city rooftops in Melbourne, Sydney and Brisbane.

Solar panels in Vatican City and Japan

Vatican City, Italy

Vatican City is the world's smallest independent state and is hoping to become the first solar- powered nation in the world. It plans to create Europe's largest solar power plant, which will provide enough energy to power all of the state's 40 000 households. The roof of the Paul VI Hall is now covered in photoelectric cells (see figure 3).

FIGURE 3 Solar panels cover the roof of the Paul VI Hall, as seen from the dome of St Peter's Basilica.

Ota, Japan

Ota is located 80 kilometres north-west of Tokyo and is one of Japan's sunniest locations. Through investment by the local government, Ota is one of Japan's first solar cities — three-quarters of the town's homes are covered by solar panels that have been distributed free of charge.

Waste incineration in Vienna

A waste incineration and heat generation plant is part of a hard-waste management system in Vienna, Austria (see figure 5). This plant became the first in the world to burn waste that cannot be recycled and use the energy generated by the plant in a heating network. The plant burns more cleanly and produces more heat and energy than many other waste generation plants, making it attractive to many urban communities. Each year, waste is turned into heat and electricity and supplied to 190 000 homes and 4200 public buildings, including Vienna's largest hospital. Landfill waste has been reduced by 60 per cent in the city.

FIGURE 4 A street in Ota, Japan — solar panels are visible on most of the houses.

FIGURE 5 Spittelau waste treatment plant in Vienna, Austria. This power station burns waste, thus reducing landfill, to produce heat that is supplied to thousands of buildings.

The Loading Dock, Baltimore

The Loading Dock is an organisation based in Baltimore, Maryland, in the United States, that recycles building material that was destined for landfill. The material is reused to help develop affordable housing while preserving the urban environment. The organisation works with non-profit housing groups, environmental organisations, local government, building contractors, manufacturers and distributors and uses human resources from within the community, improving living conditions for families, neighbourhoods and communities.

Since 1984, The Loading Dock has saved low-income housing and community projects more than US$16.5 million and has rescued over 33 000 tons of building materials from landfill. There has been interest in the project from 3000 other cities within the United States and in Mexico, the Caribbean, Hungary, Germany and five countries in Africa. All these projects will have a positive impact on people's lives and the urban environment.

9.13 Activities

To answer questions online and to receive **immediate feedback** and **sample responses** for every question, go to your learnON title at www.jacplus.com.au. *Note*: Question numbers may vary slightly.

Explain

1. The projects described in this subtopic have been completed on a local *scale*. Why do you think this is the case? Do you think any of these *sustainable* city projects would work on a suburb-wide or city-wide *scale*? Why or why not?

Think

2. Work in pairs to identify one urban problem and design a *sustainable* program that would help to improve the condition. You will need to conduct some research to find similar problems and ways in which they have been tackled. Your program should include responses to the following.
   • What is the urban problem? Include statistics (graphs or tables).
   • Where is the problem located? Describe the location and include city/state/country map/s.
   • What are the aims of your project? Describe what you hope to achieve.
   • How will you achieve your aims? Describe your program or idea.
   • Which individuals or groups are to be involved?
   • What results would reflect success for your project?

Present your program to the class in the form of a Prezi or multimedia presentation, panel discussion or other format of your choice. Alternatively, you could share your programs through a class blog or wiki.

9.14 Are there sustainable cities in Australia?

9.14.1 Our top 10

Australian cities often perform well in worldwide rankings of liveability. In one survey in 2012, four Australian cities were ranked in the top 10 — Melbourne (1), Adelaide (5), Sydney (7) and Perth (9). Liveability is an assessment of the quality of life in a particular place — living in comfortable conditions in a pleasant location. But being liveable is not the same as being sustainable, which involves living in a way that sustains the environment and conserves resources into the future.

9.14.2 Measuring city sustainability

What makes a city sustainable? In 2010, the Australian Conservation Foundation conducted a study to measure the sustainability of Australia's 20 largest cities. The indicators measured were a combination of:
• environment — air quality, ecological footprint, water, green building and biodiversity
• quality of life — health, transport, wellbeing, population density and employment
• resilience (the ability of a city to cope with future change): climate change, public participation, education, household repayments and food production.

The results showed that Darwin was the most sustainable city in Australia in 2010 (see figure 2). It performed best in terms of the economic indicators of employment and household repayments. Figure 1 shows the ranking for other cities.

FIGURE 1 A sustainability ranking of Australia's cities

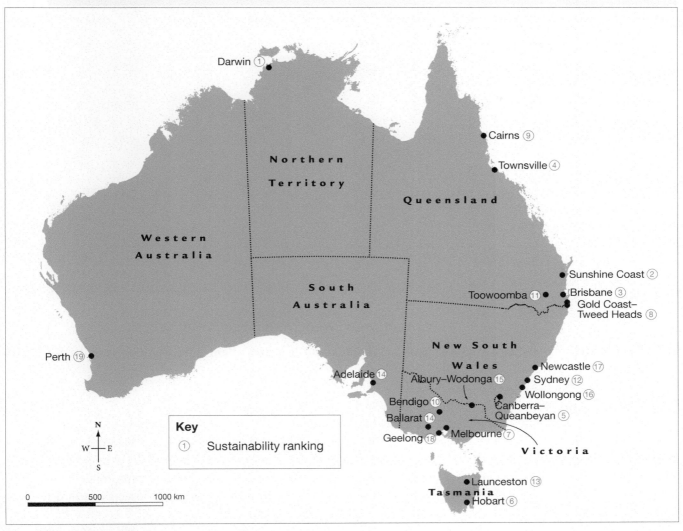

Darwin ①

Northern Territory

Cairns ⑨

Townsville ④

Queensland

Western Australia

South Australia

Sunshine Coast ②

Toowoomba ⑪ Brisbane ③

Gold Coast– Tweed Heads ⑧

Perth ⑲

Adelaide ⑭

New South Wales

Albury–Wodonga ⑮

Newcastle ⑰

Sydney ⑫

Wollongong ⑯

Bendigo ⑩

Canberra– Queanbeyan ⑤

Ballarat ⑭

Geelong ⑱ Melbourne ⑦

Victoria

Key

① Sustainability ranking

Launceston ⑬

Tasmania

Hobart ⑥

N W–E S

0   500   1000 km

Source: MAPgraphics Pty Ltd, Brisbane

FIGURE 2 Darwin, the most sustainable city in Australia in 2010

9.14.3 Local urban communities

In most cities, it is often action at a local community scale that can make the most difference in improving city sustainability. State governments and local councils have responsibility for improving complex infrastructure (for example transport and water supply) for whole cities, but change at a local level can have positive results.

Sustainable communities in cities may have some of the following in common:

- friendly and social communities
- consume less energy and water and produce less waste
- have **medium-** to **high-density** rather than **low-density housing**
- are within walking distance of some public facilities and have excellent public transport links for longer trips
- include public places that people can walk to
- have good landscaping
- dwellings have been built to a budget to make them affordable.

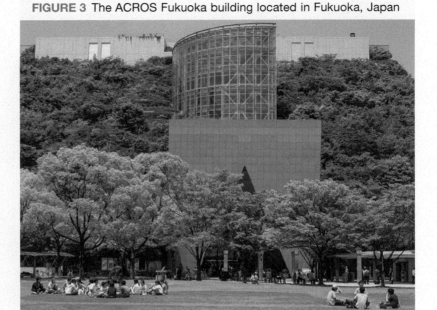

FIGURE 3 The ACROS Fukuoka building located in Fukuoka, Japan

The ACROS Fukuoka building located in Fukuoka, Japan is an example of plants and greening being used to enhance a building (figure 3). The terraced green roof and green walls merge with a park and contain around 35 000 plants. The green roof keeps the temperature inside more constant and comfortable, thus reducing energy consumption. It is also able to capture rainwater run-off and attracts many insects and birds. In addition, it is visually appealing and attracts many people to the surrounding park.

CASE STUDY

Christie Walk, Adelaide

Christie Walk is located in Adelaide in South Australia. It is a small urban village of 27 dwellings located on a quarter of an acre of land. The site is within easy walking distance of Adelaide's markets, parklands and CBD, which means car use is reduced. Around 40 people live at Christie Walk, ranging in age from very young to over 80 years.

A number of principles were used in the design of Christie Walk.

- Low energy demand (passive heating and cooling; natural lighting and sealed double glazing in all windows and glass doors)
- Maximising the use of renewable/solar-based energy sources (photovoltaic cells on the roof) and minimising the use of non-renewable energy sources
- Capturing and using storm water (in large underground rainwater tanks) and recycling waste water
- Creating healthy gardens and maximising the biodiversity of indigenous flora and fauna. The gardens also produce herbs, vegetables and fruit.
- Avoiding the use of products that damage human health
- Minimising the use of non-recyclable materials

FIGURE 4 One of the sustainable buildings in Christie Walk

FIGURE 5 A plan of Christie Walk in Adelaide

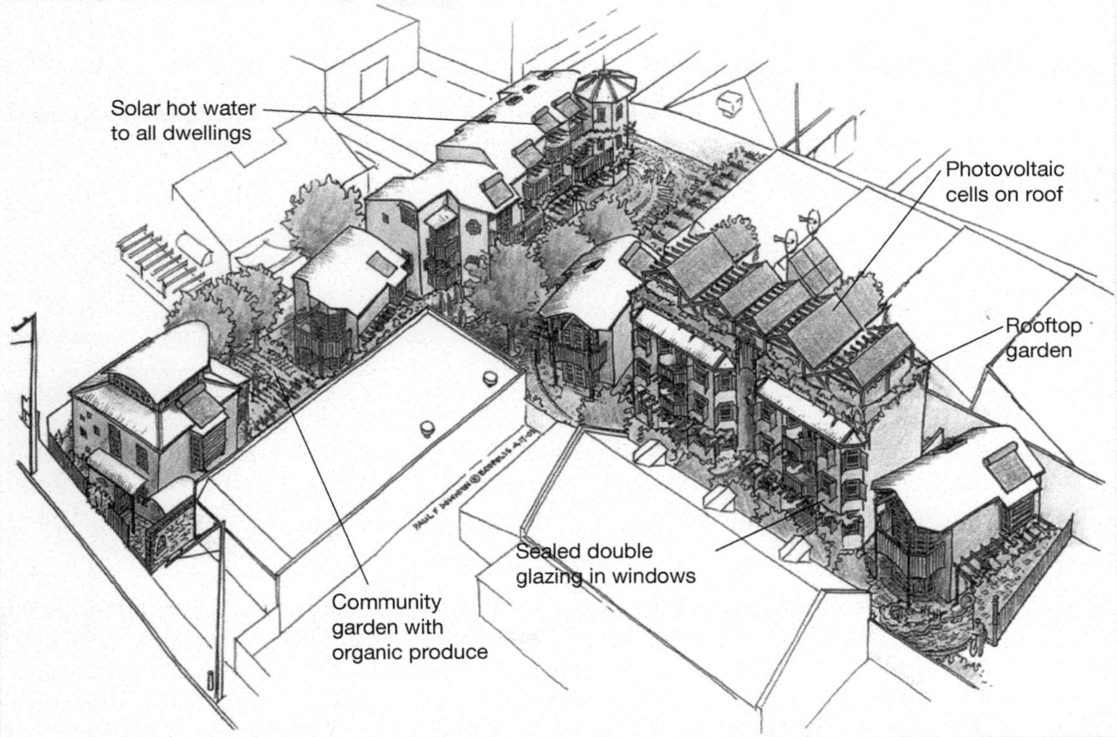

Solar hot water to all dwellings

Photovoltaic cells on roof

Rooftop garden

Sealed double glazing in windows

Community garden with organic produce

FIGURE 6 Rooftop gardens provide good insulation, protecting the buildings below from the hot sun in summer. In winter, they keep warmth from escaping from the building below.

9.14 Activities

To answer questions online and to receive **immediate feedback** and **sample responses** for every question, go to your learnON title at www.jacplus.com.au. *Note:* Question numbers may vary slightly.

Explain

1. In your own words, describe the difference between a liveable and a *sustainable* city.

Discover

2. A hectare is equivalent to 10 000 square metres, or about 2.5 acres. In urban Australia, most houses were traditionally built on quarter-acre blocks (about 12 house blocks per hectare).
    (a) Walk around your neighbourhood or school area and pace out 100 x 100 metres. This gives you an idea of what one hectare looks like.
    (b) Use Google Earth or Google Maps to count or estimate the number of dwellings in your local area.
    (c) Compare your data with the definitions for low-, medium- and high-density housing. What type of housing density is in your local area?

Predict

3. Use the **Sustainable cities index** weblink in the Resources tab to find out how cities in your state or territory have performed in measurements of *sustainability*. List five things that could improve *sustainability* in these cities. Is there anything you personally can do to make a difference?

Think

4. Work in groups of three. Use the **Sustainable Cities Awards** weblink in the Resources tab to learn about projects that have won Sustainable Cities Awards in Australia. Each group member should read about three awards and summarise the projects to the others. Using a diamond ranking chart, rank the projects from

most to least important for *sustainability*. Write the name or description of the best project in the top space of the diamond and the least *sustainable* at the bottom. Add the other projects to the chart after your group has discussed and agreed on the ranking.

5. Use ideas from this subtopic and further research to design a small *sustainable* urban neighbourhood. You may choose to work in groups or individually. You may like to use photographs of examples you find in your city/town or on the internet to draw your plan. Alternatively, video some examples and incorporate them into your design. Justify the inclusion of all the features you choose by annotating the plan or writing some notes to explain your choices. Present your final plan to the class as a panel presentation or on a class blog or wiki.

 **Deepen your understanding of this topic with related case studies and questions.**
 ❷ **Brisbane: an eco-city**

 Explore more with these weblinks: Sustainable cities index, Sustainable Cities Awards

9.15 Review

9.15.1 Review
The Review section contains a range of different questions and activities to help you revise and recall what you have learned, especially prior to a topic test.

9.15.2 Reflect
The Reflect section provides you with an opportunity to apply and extend your learning.

Access this subtopic at **www.jacplus.com.au**

TOPIC 10
Planning Australia's urban future

10.1 Overview

Numerous **videos** and **interactivities** are embedded just where you need them, at the point of learning, in your learnON title at www.jacplus.com.au. They will help you to learn the content and concepts covered in this topic.

10.1.1 Introduction

We often hear the word *sustainable*, but what does it mean? Sustainability means meeting our own current needs while still ensuring that future generations can do the same. To make this happen, human and natural systems must work together without depleting our resources. Ultimately, sustainability is about improving the quality of life for all — socially, economically, and environmentally — both now and in the future. In the words of HRH The Prince of Wales, 'Remember, our children and our grandchildren will ask not what our generation said, but what they did'.

Green walls of an 'eco office'

10.2 What do sustainable cities look like?

10.2.1 A common purpose

Our cities are facing an important challenge. Some predict that Australia's population will reach 45 million by 2050. If this is the case, then our cities must change and adapt to become more efficient in order to maintain or improve our current quality of life. How will we cope with a growing population?

Sustainable communities share a common purpose of building places where people enjoy good health and a high **quality of life**. A sustainable community can thrive without damaging the land, water, air, natural and cultural resources that support them, and ensures that future generations have the chance to do the same. The basic **infrastructure** should be designed to minimise consumption, waste, pollution and the production of greenhouse gases. Sustainable urban areas strike a delicate but achievable balance between the economic, environmental and social factors.

A sustainable city is one that has a small **ecological footprint**. The ecological footprint of a city is the surface area required to supply a city with food and other resources and to absorb its wastes. At the same time, a sustainable city is improving its quality of life in health, housing, work opportunities and liveability.

We can address the challenges and opportunities for sustainable communities at two different scales: neighbourhood and city level.

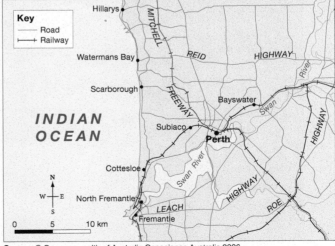

FIGURE 1 Perth, Western Australia. Building sustainable communities means we have to work at various scales.

Source: © Commonwealth of Australia Geoscience Australia 2006

FIGURE 2a An inner-city organic community urban farm in Perth, Western Australia

Ways to improve sustainability at the neighbourhood scale:
- reducing the ecological footprint
- protecting the natural environment
- increasing community wellbeing and pride in the local area

- changing behaviour patterns by providing better local options
- encouraging compact or dense living
- providing easy access to work, play and schools.

FIGURE 2b An aerial view of the Swan River and the city of Perth, Western Australia

Ways to improve sustainability at the city scale:
- building strong central activities areas (either one major hub, or a number of specified activity areas)
- reducing traffic congestion
- protecting natural systems
- avoiding suburban sprawl and reducing inefficient land use
- distributing infrastructure and transport networks equally and efficiently to provide accessible, cheap transportation options

- promoting inclusive planning and urban design
- providing better access to healthy lifestyles (e.g. cycle and walking paths)
- improving air quality and waste management
- using stormwater more efficiently
- increasing access to parks and green spaces
- reducing car dependency and increasing walkability
- promoting green space and recreational areas
- demonstrating a high mix of uses (e.g. commercial, residential and recreational).

10.2 Activities

To answer questions online and to receive **immediate feedback** and **sample responses** for every question, go to your learnON title at www.jacplus.com.au. *Note*: Question numbers may vary slightly.

Remember

1. Complete the following sentence: Some organisations have projected that Australia's population will reach __ million by _____.
2. What are the two main aims of a *sustainable* community?

Explain

3. Explain the term *ecological footprint* in your own words.
4. What are the two *scales* at which we can work to improve the sustainability of our communities? What are some of the differences between the two?

Discover

5. (a) How is an ecological footprint measured? Use the **Ecological footprint calculator** weblink in the Resources tab, or a teacher recommendation, to work through the steps to determine your own ecological footprint.
   (b) After using the calculator, compare your ecological footprint with those of your classmates by creating a continuum on the board. It should start from smallest footprint (least planets consumed) to largest footprint (most planets consumed). Discuss which areas you think contributed to the wide variety of footprints.

Think

6. What might a *sustainable* home look like to you?
7. Consider the areas listed in which a neighbourhood can become more *sustainable*. Create a table and, from your own perspective, detail the ways in which you believe your own suburb or neighbourhood is meeting these aims. Add another column and use the internet to research how your local council is trying to make your suburb more *sustainable*. Conclude by writing a few sentences to answer the following questions:
   (a) Is my neighbourhood *sustainable*?
   (b) How will liveability be improved?
   (c) What needs to *change* in order to make it even more *sustainable*?

 **learn**on RESOURCES — ONLINE ONLY

 Try out this interactivity: Ecological footprint (int-3121)

 Explore more with this weblink: Ecological footprint calculator

 my**World**Atlas **Deepen your understanding of this topic with related case studies and questions.**
   ❂ **Brisbane: an eco-city**

10.3 Are growing urban communities sustainable?

10.3.1 The urban explosion

In 2008, for the first time in history, the world's urban population outnumbered its rural population. In 2015, the world's population reached over 7.3 billion; it is expected to reach 9.2 billion by 2050. Where will all these people live? What challenges will cities and communities face in trying to ensure a decent standard of living for all of us?

Global population growth will be concentrated mainly in **urban** areas of developing countries. It is forecast that by 2030, 3.9 billion people will be living in cities of the developing world. The impact of expanding urban populations will vary from country to country and could prove a great challenge if a country is not able to produce or import sufficient food. Hunger and starvation may increase the risk of social unrest and conflict. On the other hand, farmers can help satisfy the food needs of expanding urban populations and provide an economic **livelihood** for people in the surrounding region.

One of the biggest challenges we face is ensuring that the sustainability of our economy, communities and environment is compatible with Australia's growing urban population (see table 1).

TABLE 1 Percentage of population residing in urban areas by country, 1950–2050

	1950	1975	2000	2025	2050
Australia	77.0	85.9	87.2	90.9	92.9
Brazil	36.2	60.8	81.2	87.7	90.7
Cambodia	10.2	4.4	18.6	23.8	37.6
China	11.8	17.4	35.9	65.4	77.3
France	55.2	72.9	76.9	90.7	93.3
India	17.0	21.3	27.7	37.2	51.7
Indonesia	12.4	19.3	42.0	60.3	72.1
Japan	53.4	75.7	78.6	96.3	97.6
Papua New Guinea	1.7	11.9	13.2	15.1	26.3
United Kingdom	79.0	77.7	78.7	81.8	85.9

Source: UN Population Division 2011

10.3.2 The future for Australia

Australia's population will continue to grow and change. In particular, it will become more urban and its composition will age. Population increase threatens our fragile Australian environment. We continue to witness loss of biodiversity, limits on water supply, more greenhouse gas emissions and threats to food security. Our cities experience more traffic congestion and there are problems with housing availability and **affordability**. Access to services, infrastructure and green space are limited for some people in our communities. To handle these many challenges, we must plan effectively for an increased population by building communities that can accommodate future changes. This will build communities in which all Australians live and prosper.

10.3.3 The rural lifestyle

Approximately 93 per cent of Australia's growing population will be living in urban areas by 2050 (see table 1). However, some urban residents will make a 'tree change' or a 'sea change' and relocate to rural areas or the coast. The population in rural communities is generally stable or decreasing, as many young people leave in search of jobs and study opportunities. Some rural communities manage to keep their populations stable by shifting their employment focus from manufacturing to services; by utilising better internet connections, to allow people to work remotely from their office; or by improving public transport links.

FIGURE 1 Percentage of population in urban areas, 2011–2015

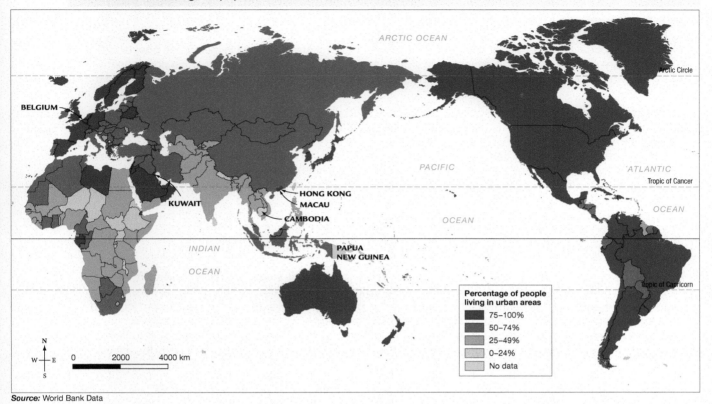

Percentage of people living in urban areas
- 75–100%
- 50–74%
- 25–49%
- 0–24%
- No data

Source: World Bank Data

FIGURE 2 Change in Australian urban and rural populations over time

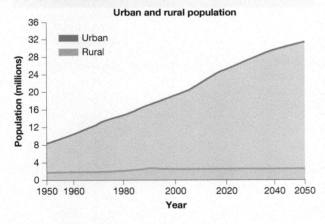

FIGURE 3 The narrowing gap between rural and urban populations, Afghanistan

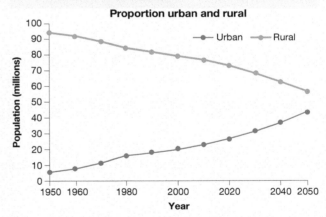

10.3 Activities

To answer questions online and to receive **immediate feedback** and **sample responses** for every question, go to your learnON title at www.jacplus.com.au. *Note:* Question numbers may vary slightly.

Remember

1. (a) The global population is *changing*. Where will most of the world's population live in the future?
   (b) Is the pattern of population *change* the same or different in Australia?

Explain

2. Refer to table 1.
   (a) Which countries will be the most and least urbanised in 2050?
   (b) Which countries are predicted to experience the greatest percentage *change* in their urban population?
   (c) Are there any countries that have not seen a gradual increase in their percentage of urban population since 1950? Why might this be the case? (You may need to conduct some additional research using the internet.)

3. Examine table 1. Create a bar graph that shows the *change* over time for four countries of your choice.

Discover

4. Growing communities create growing problems. For example, social problems may include poverty, chronic unemployment, welfare dependence, drug and alcohol abuse, crime and homelessness. Working in small groups, brainstorm some of the impacts that growing communities may have on (a) the *environment* and (b) the economy.

5. Young people leave rural areas in search of employment and education. What factors could contribute to you leaving the area where you live?

Predict

6. In cities, we must face the challenges and opportunities of productivity, *sustainability* and liveability. If we address one goal, we can have an impact, either positively or negatively, on others. This demonstrates *interconnection*. For example, efficient public transport can fix congestion and improve access to jobs and opportunity (productivity). It can also reduce greenhouse gas emissions (*sustainability*) and make access to education, health and recreational facilities more affordable (liveability). Using the example of the National Broadband Network, how might productivity, *sustainability* and liveability be affected? Classify the effects you have listed as positive or negative.

10.4 SkillBuilder: Reading and describing basic choropleth maps

WHAT IS A CHOROPLETH MAP?

A choropleth map is a shaded or coloured map that shows the density or concentration of a particular aspect of an area. The key/legend shows the value of each shading or colouring. The darkest colours usually show the highest concentration, and the lightest colours usually show the lowest concentration.

FIGURE 1 Population density in Brazil

Go online to access:

- a clear step-by-step explanation to help you master the skill
- a model of what you are aiming for
- a checklist of key aspects of the skill
- a series of questions to help you apply the skill and to check your understanding.

learn on RESOURCES — ONLINE ONLY

- **Watch this eLesson:** Reading and describing basic choropleth maps (eles-1706)
- **Try out this interactivity:** Reading and describing basic choropleth maps (int-3286)

10.5 Should we manage our suburbs?

10.5.1 Living on the edge

There is much at stake on the rural–urban fringe, with the conflict between farming and urban residential development reaching a critical point on the outskirts of Australia's cities. Australia is the driest inhabited continent on Earth, and just six per cent of its total area (45 million hectares) is arable land. The areas targeted by our state governments for residential development continue to expand. When some of our most fertile farmland is lost to **urban sprawl**, we reduce our productive capacity. Is this a recipe for sustainability?

On the edge of many Australian cities, new homes are being built as part of planned developments on greenfield sites. These were previously **green wedges**, wildlife habitats and productive farmland on the urban fringe. Accompanying these housing developments are plans for kindergartens, schools, parks, pools, cafés and shopping centres (often called amenities and facilities).

Having an 'affordable lifestyle' is the main attraction for people who purchase these brand new homes. They like the idea of joining a community and having the feeling of safety in their newly established neighbourhood.

Most new houses on the rural–urban fringe are bought by young first-home buyers, attracted by cheaper housing and greener surroundings. Generally the residents of these fringe households feel that the benefits of their location outweigh the poor public transport provisions and long journeys to work and activities — trips that are usually made in a car.

FIGURE 1 The history of Sydney's urban sprawl

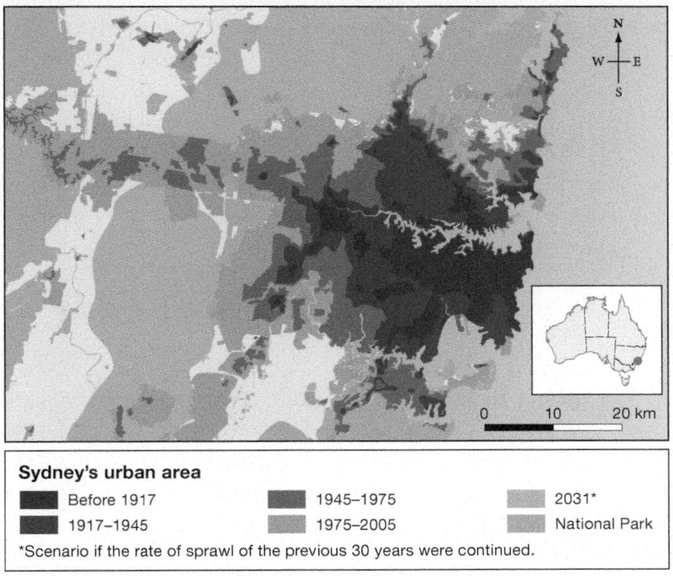

Sydney's urban area
- Before 1917
- 1917–1945
- 1945–1975
- 1975–2005
- 2031*
- National Park

*Scenario if the rate of sprawl of the previous 30 years were continued.

Source: Provided by Metropolitan Strategy, NSW Department of Planning & Infrastructure

10.5.2 Feeding our growing cities

Market gardens have traditionally provided much of a growing city's food needs, supplying produce to central fruit and vegetable markets. These 'urban farms' were located on fertile land within a city's boundaries but close to its edge, with a water source nearby and often on floodplains. They have been in existence in and around Australia's major cities since the 1800s, and some (such as Burnley Gardens in Richmond, Victoria) are now listed on the National Trust heritage garden register.

Fifty per cent of Victoria's fresh vegetable production still occurs in and around Melbourne, on farms such as those at Werribee and Bacchus Marsh. More than 60 per cent of Sydney's fresh produce is grown close to the city, with the bulk of it coming from commercial gardens such as those in Bilpin, Marsden Park and Liverpool.

FIGURE 2 The battle at our urban fringe: housing or farmland?

These farms are important because:

- they provide us with nutritious food that does not have to be transported very far
- they provide local employment
- they preserve a mix of different land uses in and around our cities.

Currently, we can obtain our food from almost anywhere because we have modern transportation (such as trucks and planes), better storage technology (refrigeration and ripening techniques) and cheap sources (not necessarily the closest). However, this fails to recognise that Australia's population may double by 2050 and food will become more scarce on a global level. The eradication of our local food providers may be at our own peril.

Land use zoning is generally the responsibility of state planning departments but cooperation is required by all three levels of government: local, state and federal. We need to ensure that our green wedges are protected from becoming **development corridors**. The needs on both sides of the argument are valid. How can we house a growing population and provide enough food for them? Can we do both?

10.5 Activities

To answer questions online and to receive **immediate feedback** and **sample responses** for every question, go to your learnON title at www.jacplus.com.au. *Note*: Question numbers may vary slightly.

Remember

1. List the groups involved in the conflict over our rural–urban *spaces*.

Explain

2. Why it is important for people to have rural *spaces*, such as market gardens, close to the city?
3. Why it is important for cities to have access to more land for urban development?
4. Refer to figure 1. Describe how Sydney's urban sprawl has *changed* in direction and pace.

Discover

5. Use the internet to research some companies that sell house and land packages in your state. What are some of the marketing messages that are used to sell the properties? Do you think they are able to deliver on their promises?
6. Many new homes on the urban fringe are built with six-star or seven-star energy efficiency. Use the internet to help you find out what this means.
7. Between 2012 and 2015, seven new suburbs were added in the Melbourne metropolitan area. They were Diggers Rest, Lockerbie, Lockerbie North, Manor Lakes, Merrifield West, Rockbank and Rockbank North. Using a mapping website such as Google Earth, locate these *places* using a 'pin' on a map of Melbourne. Use a ruler tool to measure the distance from each suburb to Melbourne's CBD. Alternatively, you may be able to complete this activity in your own state by researching new suburbs added to your city.

Think

8. Housing and agriculture demands on land are two of the biggest dilemmas of the twenty-first century. A growing population needs to be housed, but it also needs to be fed, and the cost of relying on imported food can be very high. Set up a debate with your classmates on the following statement: 'Green belts close to the city should be preserved and protected.' The affirmative team will argue for this, while the negative team will argue that green belts should be removed and used for new housing developments.
9. 'Sprawl is created by people escaping sprawl.' Discuss this statement with a small group.

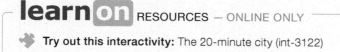 **learn ON** RESOURCES — ONLINE ONLY

  Try out this interactivity: The 20-minute city (int-3122)

10.6 How can we manage traffic?

10.6.1 The way forward

How did you get to school today? How long did you spend in the car? Were you stuck in a traffic jam? Australians who live in cities are experiencing longer commuting times than ever before, and this is only going to get worse. A growing population will mean an increase in cars — unless we start to tackle the problem from a sustainable perspective.

Transport is one of the largest sources of greenhouse gas emissions in Australia (34 per cent), with passenger cars contributing more greenhouse gases than any other part of the transport sector. Some of the big issues in improving the sustainability of our transport systems are listed below.

Improving our infrastructure

Better public transport infrastructure will help improve the sustainability of our communities. Some cities have excellent rail systems or electrified tramways that were installed many years ago. But as cities grow and change, costly extensions may be required. Buses are much cheaper and quicker to upgrade. In Curitiba, Brazil, **bi-articulated buses** travel in dedicated bus lanes, and 70 per cent of the population uses the service. Public transport systems are cost-effective because it costs the same to run a bus or train with one passenger as it does with 1000 passengers (see figure 1). The more people who travel, the less it costs to transport each person.

Technologically advanced transportation

Since the late twentieth century, there have been many improvements in car design, occupant safety and fuel efficiency. China hopes to sell 500 000 hybrid or electric cars annually by 2015; Paris, London and Sydney have started installing electric car charging stations around the city and car companies like Tesla only produce electric-powered cars. Sustainable public transportation methods, however, have not attracted the same interest or **investment**. Adelaide and the Canadian town of Whistler have led the way with sustainable public transport options. In Adelaide, the Tindo Bus is powered by solar energy and can run 200 kilometres between charges. Whistler has the largest fleet of hydrogen fuel cell buses which emit only water vapour.

Denser urban settlements

When an urban area is dense, the buildings are more compact, and more people live there. Dense urban settlements have 'efficiencies' already built in. Older cities, such as those in Europe, were established long before the invention of motor vehicles, meaning that they were built for walking. The older parts of European cities have narrow streets and laneways, and cannot

FIGURE 1 Greenhouse gas emissions from different forms of transport

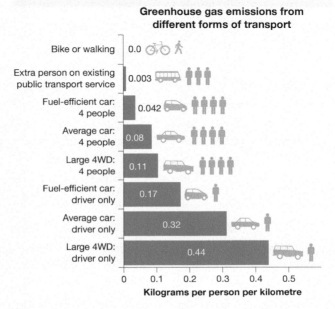

Greenhouse gas emissions from different forms of transport

	Kilograms per person per kilometre
Bike or walking	0.0
Extra person on existing public transport service	0.003
Fuel-efficient car: 4 people	0.042
Average car: 4 people	0.08
Large 4WD: 4 people	0.11
Fuel-efficient car: driver only	0.17
Average car: driver only	0.32
Large 4WD: driver only	0.44

FIGURE 2 The Tindo bus in Adelaide runs on solar energy.

cope with congestion. Europeans are less likely to own cars because they live close to their daily destinations, and this reduces the need for cars. In Manhattan, New York, 82 per cent of people (1.3 million) travel to work by public transport, bicycle or foot — this is 10 times the rate of the average American.

10.6.2 Changing our behaviour

Did you use a sustainable form of transport to get to school today? Cycling and walking are forms of mass urban transportation. Providing safe bike paths and

FIGURE 3 Primary students catch the walking school bus

walking routes makes people more likely to change their behaviours. If you have to travel by car, one way of increasing the effectiveness of each trip is car pooling. Governments or workplaces may also provide **incentives** for individuals to make a more sustainable transport choice.

Positive changes are happening, even if it is a little slow. The most recent figures show that 2.3 per cent of Victorians rode their bicycles to work and 4.1 per cent walked to work in 2012 compared to 1.1 per cent and 2.9 per cent respectively in 2001.

The toll we pay

Travel, particularly in our own cars, has increased at a rapid rate over the past 50 years. We have increased our mobility, independence and opportunities, and this has transformed the way in which land is used and people live. But as well as these benefits, car travel has created many health problems. Accidents and injury, climate change, air, water, soil and noise pollution, reduction in social interaction, and declining physical activity are all negative effects of car travel that take their toll on our health.

10.6 Activities

To answer questions online and to receive **immediate feedback** and **sample responses** for every question, go to your learnON title at www.jacplus.com.au. *Note*: Question numbers may vary slightly.

Remember

1. Study figure 1. Which mode of transportation contributes the most greenhouse gas emissions and which contributes the least?

Explain

2. What mode of transportation did you use to come to school today? How long did it take? How did your family members travel to their *place* of work or their school or university today? Use an internet mapping tool to help you work out how many kilometres your family travelled and by what means.

Discover

3. Tally the results for your class's responses to question 2. Present the information in graph format. If possible, compare your results with another class.
4. What is car pooling? As a class, work out the minimum number of cars it would take to efficiently transport your entire class to school.
5. Create a mind map of the way car travel affects your health, and then create a corresponding mind map of the way public transport affects your health. Include as many positive and negative points as you can with a brief explanation.

Predict

6. Consider the four areas for improvement listed in this subtopic. Which do you think will be the most important for (a) individuals and (b) the government to focus on in the next five years?
7. Download a map of your suburb and print it out. Annotate it with current public transport options, such as trains, buses, bike paths and footpaths. Use different colours and a key to suggest improvements to existing options in your local *space*.

10.7 SkillBuilder: Drawing a line graph using Excel

WHAT IS A LINE GRAPH?

A line graph is a clear method of displaying information so it can be easily understood. Using a digital means of drawing a line graph enables you to show multiple data sets clearly.

Go online to access:

- a clear step-by-step explanation to help you master the skill
- a model of what you are aiming for
- a checklist of key aspects of the skill
- a series of questions to help you apply the skill and to check your understanding.

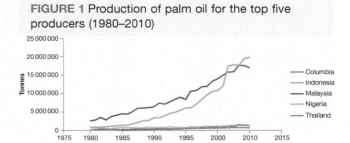

FIGURE 1 Production of palm oil for the top five producers (1980–2010)

SOURCE: FOOD and Agriculture Organization of the United Nations 2012 FAOSTAT, http://faostat3.fao.org/home//index.html

10.8 Welcome to Sustainaville

10.8.1 Why don't we just build more roads?

In an ideal world, a sustainable transport system would have a fast, clean, reliable and regular train service with waiting times of no more than ten minutes, day or night. Trams and buses would link into the train network, bringing people to the main parts of the system. Trams and buses would have priority over other traffic and run on the weekends. Station staff would be present at all times and the services would be safe and clean. What are some of the costs, other than financial, of using our cars instead of public transport?

Contrary to popular belief, building new roads and freeways does not actually ease **congestion**. This is because a new road simply becomes an opportunity for people to make new journeys that they may not have contemplated before; or they make the same journey more often; or they drive instead of taking public transport; or travel longer distances to accomplish the same task. All these things result in increased traffic on the new road, so the road system ends up just as congested as before. More energy and resources are consumed, and more pollution is generated.

FIGURE 1 Traffic jams slow down people and the economy.

FIGURE 2 This cyclist in China may be wearing a mask to reduce the effects of air pollution.

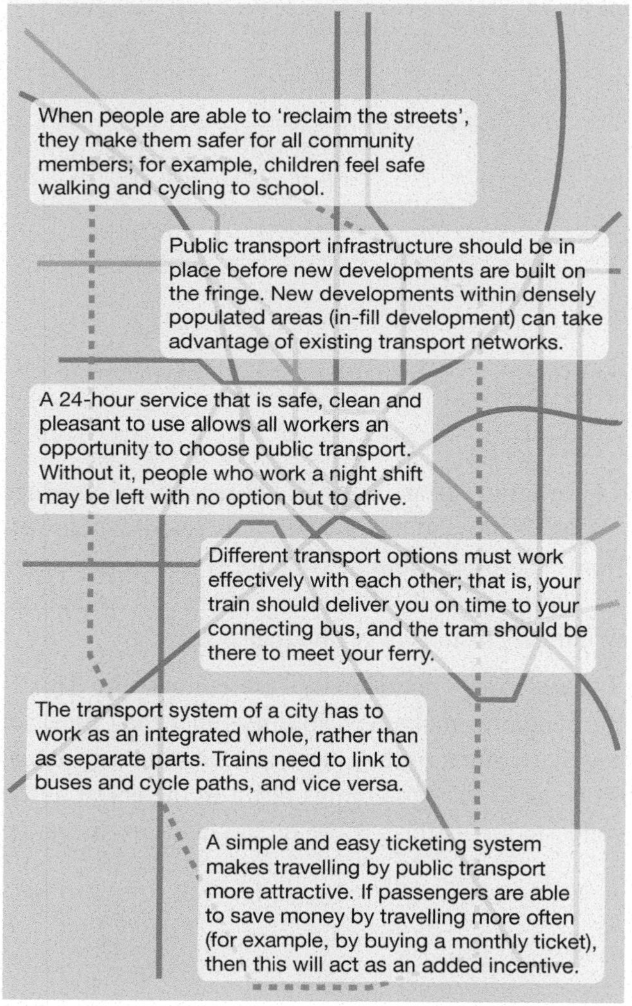

FIGURE 3 A model for transport in a sustainable city

When people are able to 'reclaim the streets', they make them safer for all community members; for example, children feel safe walking and cycling to school.

Public transport infrastructure should be in place before new developments are built on the fringe. New developments within densely populated areas (in-fill development) can take advantage of existing transport networks.

A 24-hour service that is safe, clean and pleasant to use allows all workers an opportunity to choose public transport. Without it, people who work a night shift may be left with no option but to drive.

Different transport options must work effectively with each other; that is, your train should deliver you on time to your connecting bus, and the tram should be there to meet your ferry.

The transport system of a city has to work as an integrated whole, rather than as separate parts. Trains need to link to buses and cycle paths, and vice versa.

A simple and easy ticketing system makes travelling by public transport more attractive. If passengers are able to save money by travelling more often (for example, by buying a monthly ticket), then this will act as an added incentive.

10.8.2 The benefits of an efficient public transport system

By shifting from car trips to public transport we can improve our **triple bottom line**. In other words, we improve economic efficiency, help the natural environment and do something good for society.

However, we also know that people will not get out of their cars and use public transport until public transport offers a high-quality, convenient and affordable service. Australia needs to make huge improvements in service frequency, connections and coverage. This formula has worked in other cities around the world and could work here in Australia.

Here in Australia we must look to develop Sustainaville — a community with its focus on public transport, walking and cycling.

10.8 Activities

To answer questions online and to receive **immediate feedback** and **sample responses** for every question, go to your learnON title at www.jacplus.com.au. *Note*: Question numbers may vary slightly.

Remember

1. Study figure 3. What does a public transport system need to be like in order to be a success?
2. Which three areas does the triple bottom line concern?

Explain

3. The benefits of an efficient public transport system are many. If we were to discuss its impact on the *environment*, we would see less air and noise pollution, conservation of green spaces (public transport uses less space than roads), and reduced greenhouse gas emissions (GHGE). A full train produces about five times less GHGE than the cars needed to move the same number of people. Explain how an efficient public transport system would benefit the economy and society, following the example above to assist you.

Discover

4. Curitiba in Brazil has installed a very successful bus rapid transit system (BRT), which has buses running about every 90 seconds and is used by 70 per cent of Curitiba's residents. Conduct some internet research using the **Urban habitat** weblink in the Resources tab or other sites, or view one of the many videos available online about the BRT system. Make a list of the unique features of the BRT and include some facts about the effect the system has had on the triple bottom line of Curitiba. How does this system compare to those you are aware of in your local community here in Australia?

Predict

5. Use the **Crank busters** and **Transport urban myths** weblinks in the Resources tab to find information that will help you create a 'True or false?' quiz about public transport for your classmates.

Think

6. What kind of public transport system would you like to use? Design your own regional public transport option, using your local council area borders. Create a brochure showcasing the many benefits and features of the service. Include a map that details the routes of the service, frequency of service, hours of operation and cost. Use figure 3 to assist you.
7. There are many arguments for getting out of our cars and onto trams, trains, buses or bikes. Use the **Crank busters** and **Transport urban myths** weblinks in the Resources tab and other resources to prepare a class debate on one of the following topics:
   - People who own cars won't use public transport.
   - Bringing back tram conductors and station staff would increase fares.
   - Cars are more efficient than public transport.
   - Freeways reduce traffic congestion and pollution.
   You may be able to share the topics listed above among different groups and then present to the entire class.

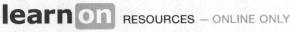 **RESOURCES** — ONLINE ONLY

   🔗 **Explore more with these weblinks:** Urban habitat, Crank busters, Transport urban myths

10.9 Where are the world's sustainable cities?

Access this subtopic at **www.jacplus.com.au**

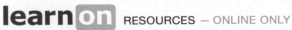 **RESOURCES** — ONLINE ONLY

   📄 **Complete this digital doc:** Masdar City infographic (doc-11470)

10.10 Can we plan to 'live vertically'?

Australian cities are experiencing an apartment revolution. More people are choosing to live in the centre of cities in high-rise apartments rather than in houses on big suburban blocks. Urban life now sees families and individuals moving to the inner city for a variety of reasons, such as seeking to make a smaller ecological footprint, or avoiding long commutes to school, work and shops.

10.10.1 Higher-density living, smaller households

Australian households are changing in structure all the time, and recent data suggests the greatest increase will be in **family households**, which will grow from 5.6 million in 2006 to 8 million households in 2031. Family households are projected to remain the most common type in Australia. Although they show the greatest increase in numbers, single-person households are projected to experience the greatest percentage increase — 63 per cent — over the next 25 years, from 1.9 million in 2006 to 3 million in 2031. This is due to the ageing of Australia's population and the fact that older women are more likely to live alone than men. It will be a challenge to provide enough accommodation and make residences as sustainable as possible.

FIGURE 1 Projected number and type of households in Australia, 2006–2031

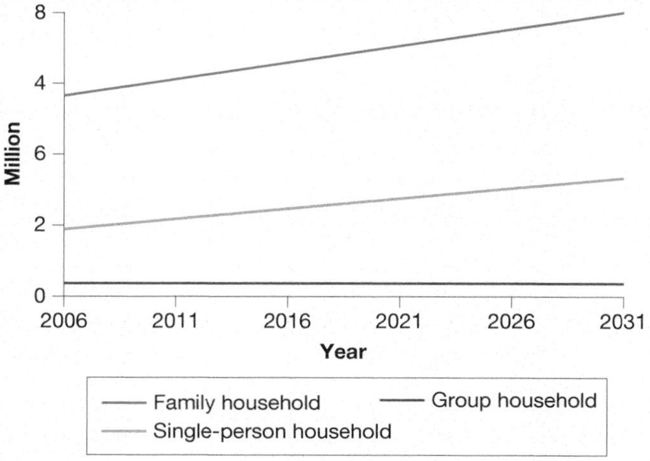

10.10.2 Going green

All housing can be designed to be sustainable. However, medium- and higher-density housing can offer the greatest opportunity for energy savings. Buildings with shared walls and more than one storey (such as two-storey and semi-detached homes, terraces and apartments) use less energy for heating and cooling than single-storey detached homes.

In Australia, people have started to value being able to walk to facilities and workplaces, so our urban centres are increasing in population density. For business and residential purposes, urban sprawl is far less sustainable than high-rise buildings. A sustainable building may include on-site energy generation (such as solar panels and wind turbines) and passive energy design (such as insulation), reducing the need for air-conditioning and heating. 'Green' or recycled building materials can also lower the environmental costs of construction.

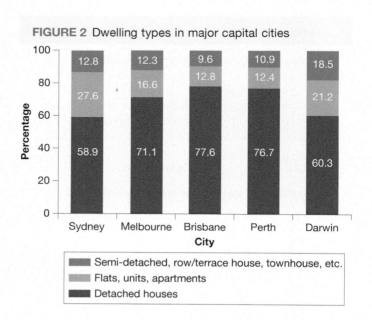

FIGURE 2 Dwelling types in major capital cities

Green roofs and walls

Green roofs and walls have a history dating back thousands of years. People are rediscovering the benefits of creating healthy, green buildings. A green, or living, roof is a roof surface that is planted partially or completely with vegetation over a waterproof layer. They may be extensive, with simple ground-cover vegetation, or intensive, with soil more than 200 millimetres deep and planted with trees. Green walls are external or internal walls of buildings that include vegetation, either in stacked pots or in growing mats.

Green roofs are now an accepted part of modern buildings in Europe. Approximately 10 per cent of German roofs have been greened, and the city of Linz, in Austria, requires green roofs on all new residential and commercial buildings with rooftops over 100 square metres. (To see how big this is, pace out an area 10 metres long by 10 metres wide.)

Green roofs have several benefits. They:
- are aesthetically pleasing
- provide a cooling effect on local microclimate
- reduce carbon dioxide (CO_2)
- reduce air pollution
- provide insulation for buildings
- provide recreational space for local residents and workers.

FIGURE 3 The ACROS Building in Fukuoka, Japan, has a fully grown forest on one terraced roof.

The high life

In the last century, Europe has transformed itself from a largely rural to a mostly urban continent. It is estimated that around 70 per cent of the EU population — approximately 350 million people — lives in urban centres of more than 5000 inhabitants. About two-thirds of energy demand is linked to urban consumption and up to 70 per cent of CO_2 emissions are generated in cities. The urban way of life is both part of the problem and part of the solution. The density of urban areas allows for more energy-efficient forms of housing, transport and other services. Consequently, measures to address climate change may be more efficient and cost-effective in big, compact cities than in less densely built spaces.

10.10 Activities

To answer questions online and to receive **immediate feedback** and **sample responses** for every question, go to your learnON title at www.jacplus.com.au. *Note:* Question numbers may vary slightly.

Remember

1. Study figure 1. How are Australian households predicted to *change* over the next 20 years? What type of household do you live in?
2. What type of dwelling is your residence?

Explain

3. Explain why the types of households are going to *change* in the next 20 years in Australia.

Discover

4. Using a program such as Google Earth, visit Linz in Austria. Can you locate any green roofs or other green spaces? Conduct a flyover of your capital city. How many green roofs can you find in the central business district?

Think

5. As a teenager, what do you think are some of the advantages and disadvantages of living in a high-rise or apartment building?
6. Green roofs can be built anywhere. Select a rooftop on a building at your school, and create a plan for your own green roof. To find inspiration, conduct research on successful green roofs around the world. You will need to include a design, information on size and materials needed, and how and why it would be accessed. Present your design using a program such as Prezi or PowerPoint.

10.11 Is Auroville a sustainable community?

Access this subtopic at **www.jacplus.com.au**

10.12 How do we plan for a liveable future?

10.12.1 The role of governments

Managing and planning Australia's future urban areas will take the efforts of many. We, as citizens of Australia and the world, must be prepared to make significant changes to the way we live if we wish to enjoy a good quality of life in the future. Sustainability and liveability must be on the agenda for governments, communities and individuals.

Governments can commit to sustainability in a number of ways. They may offer incentives such as **rebates** on solar panels or water-efficient showerheads. They can fund research into sustainable technologies. Governments can adopt strict planning

FIGURE 1 In Brunswick, Victoria, an old garbage dump was converted into the Centre for Education and Research into Environmental Strategies (CERES), comprising a community garden and resource and environment education centre.

regulations and well-defined urban growth boundaries. They can have clear policies on levels of air quality, business sustainability, and the construction or **retrofitting** for sustainability of 'green' buildings. They can develop land-use plans that encourage sustainability and biodiversity.

10.12.2 The role of communities

Communities and organisations are working with governments, businesses and individuals to respond to global challenges such as climate change. There are many measures in place to improve transport and mobility, develop effective use of our land, and plan and develop appropriate policies.

Communities maintain and improve infrastructure and open spaces, and can help us work at the neighbourhood level to build a more sustainable community. An example of this is the Sustainability Street program run by many councils, where residents are encouraged to work together with their neighbours on improving local liveability. They might establish community gardens or purchase solar systems in bulk. Some great examples of communities working with governments to improve liveability and sustainability are shown in figures 1, 2 and 3.

FIGURE 2 A disused railway in New York was converted into the High Line public park.

FIGURE 3 A vertical garden disguises a five-storey car park at Southbank, Victoria.

10.12.3 The role of the individual

We can all seek to enjoy a quality of life that does not damage the environment. Although you might feel powerless, in the next decade you will be making your own contribution to society and thinking about what kind of world you would like to grow old in. You will need to consider your sustainable choices in the action areas shown below. What is *your* personal sustainability plan? Ultimately, if you want to improve your quality of life and the environment, make your choices sustainable ones. You could get involved by:
- riding or walking to school each day
- establishing an eco-classroom at your school
- learning more about the connections between Aboriginal and Torres Strait Islander peoples and their land
- installing solar hot water or solar panels at your residence
- growing your own food.

FIGURE 4 Action areas

Energy

Aboriginal and Torres Strait Islander knowledge

Waste

Community

Transport

Sustainable purchasing

10.12 Activities

To answer questions online and to receive **immediate feedback** and **sample responses** for every question, go to your learnON title at www.jacplus.com.au. *Note*: Question numbers may vary slightly.

Remember

1. Who are the three key groups making our urban areas more sustainable?

Explain

2. Study figures 1, 2, 3 and 4.
   (a) What are some ways in which governments can make changes to create a more liveable future?
   (b) What are some ways in which you, as a high-school student, can make changes to create a more liveable future?
3. Use the internet to find out how a building can be made more *sustainable*.

Discover

4. Use the internet to find information about the Science and Engineering Building, also known as the Cube, at the Queensland University of Technology in Brisbane. Explore some of the ways in which the building works, not only in conserving resources but also in improving the wellbeing of its workers.

(a) Which feature of the building is most interesting to you? Why? How does it work?

(b) Could this feature, or any others, be applied to your home? Would any of the features be more suited to larger buildings, such as your school?

(c) Would you like to work in a building like this? Why or why not? Justify your explanation.

5. Research the ways in which your local council is working at a local level to improve *sustainability*. Most councils have a section on their website dedicated to actions for *sustainability*. Work in a small group to create a short presentation on the various programs at work. What kind of programs can individuals participate in?

Predict

6. As you get older, your needs, wants and priorities will *change*. Imagine you have now completed Year 12 and are ready to move out into your first share house. In a small group (representing your new housemates), agree on a list of 10 ways that you and your housemates could live more sustainably.

Think

7. Make your own personal *sustainability* plan, using a mind map to help categorise your ideas. Consider how you could make changes in various areas of your life (school, home, sport, hobbies). List the actions that you would take, and identify what the outcome would be. For example, 'I could ride to soccer practice after school instead of being driven'. Outcome: reduced GHGE from family car.

10.13 Review

10.13.1 Review

The Review section contains a range of different questions and activities to help you revise and recall what you have learned, especially prior to a topic test.

10.13.2 Reflect

The Reflect section provides you with an opportunity to apply and extend your learning.

Access this subtopic at **www.jacplus.com.au**

TOPIC 11
Geographical inquiry: Investigating an Asian megacity

11.1 Overview

Numerous **videos** and **interactivities** are embedded just where you need them, at the point of learning, in your learnON title at www.jacplus.com.au. They will help you to learn the content and concepts covered in this topic.

11.1.1 Scenario and your task

The latest liveability report for Asian megacities has been released, and residents are concerned. Populations are increasing by between one and five per cent every year, putting city infrastructure under extreme pressure.

City authorities have commissioned your team to put together a website increasing awareness of the characteristics of the Asian megacity and informing residents of current and newly proposed sustainable development planning initiatives.

Your task

Your team has been put in charge of creating a website designed to inform the residents of an Asian megacity about its characteristics. Each city will be different depending on its location, wealth or poverty, size and climate. Your investigations need to ensure that the audience can gain a comprehensive understanding of both population characteristics and city characteristics, and that any urban problems are presented. A key feature of your website will be to cover any urban solutions and innovations that are currently being implemented in your megacity.

11.2 Process

11.2.1 Process

- You can complete this project individually or invite members of your class to form a group.
- **Planning:** You will need to research the characteristics of your chosen Asian megacity. Research topics that have been loaded in the Resources tab to provide a framework for your research include: location and city characteristics (main economy, tourism, culture); population characteristics (migrants and migration, languages, religion); and urban problems, solutions and innovations. Choose a number of these topics to include in your website and ensure you add your own. Divide the research tasks among the members of your group.

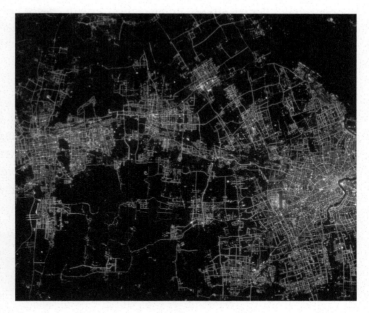

11.2.2 Collecting and recording data

Begin by discussing with your group what you might already know about your chosen Asian megacity. Then discuss the information you will be looking for and where you might find it. To discover extra information about life in your Asian megacity, find at least three sources other than the textbook. At least one of these should be an offline source such as a book or an encyclopaedia. Remember that you will need to choose specific keywords to enter into your search engine to find other data. You can view and comment on other group members' articles and rate the information they have entered.

11.2.3 Analysing your information and data

- You now need to decide what information to include in your website. Maps to show location, graphs, tables and lists to illustrate data, and images and photos with annotations (descriptive notes) should all be included. Each of these should also have a written description. You should make sure that you have addressed each of the following points.
  1. Describe the pattern of distribution on each of the maps or satellite images you have drawn or collected.
  2. What are the main characteristics of your city?
  3. How has your city changed over time? Is information available on how it is predicted to change in the future?
  4. For what reasons are people attracted to move to this city?
  5. What are the main problems in this city? Are there any solutions being introduced to try to overcome these problems?
- Download the website model and website-planning template to help you build your website.
- Use the website-planning template to create design specifications for your site. You should have a home page and at least three link pages per topic. You might want to insert features such as 'Amazing facts' and 'Did you know?' into your interactive website. Remember the three-click rule in web design — you should be able to get anywhere in a website (including back to the homepage) with a maximum of three clicks.

11.2.4 Communicating your findings

Use website-building software to build your website. Remember that less is more with website design. Your mission is to inform people about your Asian megacity in an informative and engaging way. You want people to take the time to read your entire website.

11.3 Review

11.3.1 Reflecting on your work

Think back over how well you worked with your partner or group on the various tasks for this inquiry. Determine strengths and weaknesses and recommend changes you would make if you were to repeat the exercise. Identify one area where you were pleased with your performance, and an area where you would like to improve. Write two sentences outlining how you might be able to do this.

Print out your Research Report and hand it in with your website and reflection notes.

GLOSSARY

affordability: the quality of being affordable — priced so that people can buy an item without inconvenience

altitude: height above sea level

aquifer: a body of permeable rock below the Earth's surface that contains water, known as groundwater

archaeological: concerning the study of past civilisations and cultures by examining the evidence left behind, such as graves, tools, weapons, buildings and pottery

avalanche: a sudden downhill movement of material, especially snow and ice

backwash: the movement of water from a broken wave as it runs down a beach returning to the ocean

barge: a long flat-bottomed boat used for transporting goods

basin: drainage basin; the total area drained by a river and its tributaries

bi-articulated buses: an extension of an articulated bus, with three passenger sections instead of two

blizzards: a strong and very cold wind containing particles of ice and snow that have been whipped up from the ground

catchment: area of land that drains into a river

clearfelling: a forestry practice in which most or all trees and forested areas are cut down

compost: a mixture of various types of decaying organic matter such as dung and dead leaves

congestion: the state of being overfilled or overcrowded

constructive wave: a gentle backwash that leads to material being deposited on land

convection current: a current created when a fluid is heated, making it less dense, and causing it to rise through surrounding fluid and to sink if it is cooled. A steady source of heat can start a continuous current flow.

convergent plate: a tectonic boundary where two plates are moving towards each other

coral atoll: a coral reef that partially or completely encircles a lagoon

country: the area of land, river and sea that is the traditional land of each Aboriginal language group or community; the place where they live

cultural: relating to the ideas, customs and social behaviour of a society

deposition: the laying down of material carried by rivers, wind, ice and ocean currents or waves

destructive wave: a large powerful storm wave that has a strong backwash

development corridors: area set aside for urban growth or development

divergent plate: a tectonic boundary where two plates are moving away from each other and new continental crust is forming from magma that rises to the Earth's surface between the two

downstream: nearer the mouth of a river, or going in the same direction as the current

drainage basin: an area of land that feeds a river with water; or the whole area of land drained by a river and its tributaries

ecological footprint: the amount of productive land needed on average by each person in a selected area for food, water, transport, housing and waste management

ecosystem: an interconnected community of plants, animals and other organisms that depend on each other and on the non-living things in their environment

ecotourist: a tourist who travels to threatened ecosystems in order to help preserve them

erosion: the wearing away and removal of soil and rock by natural elements, such as wind and water, and by human activity

escarpment: a steep slope or long cliff formed by erosion or vertical movement of the Earth's crust along a fault line

ethnic minority: a group that has different national or cultural traditions from the main population

EU: European Union — an economic and political union of 28 member states, mostly in Europe

evapotranspiration: the process by which water is transferred to the atmosphere from surfaces such as the soil and plants

family households: two or more persons, one of whom is at least 15 years of age, who are related by blood, marriage (registered or de facto), adoption, step-relationship or fostering

fault: an area on the Earth's surface that has a fracture, along which the rocks have been displaced

field sketch: a diagram with geographical features labelled or annotated

flash flood: a flood that occurs very quickly, often without advance warning

fly-in, fly-out (FIFO): system in which workers fly to work in places such as remote mines and after a week or more fly back to their home elsewhere

focus: the point where the sudden movement of an earthquake begins

food miles: the distance food is transported from the time it is produced until it reaches the consumer

geographical factors: reasons for spatial patterns, including patterns noticeable in the landscape, topography, climate and population

geothermal energy: energy derived from the heat in the Earth's interior

glacier: a large body of ice, formed by an accumulation of snow, which flows downhill under the pressure of its own weight

gorge: narrow valley with steep rocky walls

green wedges: also known as a green belt, this land is largely undeveloped, wild, or agricultural, and surrounds or adjoins urban areas. It prevents development of the area, allowing wildlife and flora to return or become established.

groundwater: water that seeps into soil and gaps in rocks

habitat: the total environment where a particular plant or animal lives, including shelter, access to food and water, and all of the right conditions for breeding

high-density housing: residential developments with more than 50 dwellings per hectare

host: an organism that supports another organism

hotspot: an area on the Earth's surface where the crust is quite thin, and volcanic activity can sometimes occur, even though it is not at a plate margin

human features: structures built by people

humidity: the amount of water vapour in the atmosphere

hunter–gatherers: people who collect wild plants and hunt wild animals rather than obtaining their food by growing crops or keeping domestic livestock

hydroelectric dam: a dam that harnesses the energy of falling or flowing water to generate electricity

ice age: historical period during which the Earth is colder, glaciers and ice sheets expand and sea levels fall

incentives: something that motivates or encourages a person to do something

indigenous: native to or belonging to a particular region or country

indigenous peoples: the descendants of those who inhabited a country or region before people of different cultures or ethnic origins colonised the area

infrastructure: the facilities, services and installations needed for a society to function, such as transportation and communications systems, water and power lines

intermittent: describes a stream that does not always flow

intermittent creeks: a creek that flows for only part of the year following rainfall

investment: an item that is purchased or has money dedicated to it with the hope that it will generate income or be worth more in the future

islet: a very small island

katabatic winds: very strong winds that blow downhill

lagoon: a shallow body of water separated by islands or reefs from a larger body of water, such as a sea

landslide: a rapid movement of rocks, soil and vegetation down a slope, sometimes caused by an earthquake or by excessive rain

leaching: a process that occurs in areas of high rainfall, where water runs through the soil, dissolving minerals and carrying them into the subsoil. The process can be compared to a coffee pot in which the water drips through the coffee grounds.

liquefaction: transformation of soil into a fluid, which occurs when vibrations created by an earthquake, or water pressure in a soil mass, cause the soil particles to lose contact with one another and become unstable. For this to happen, the spaces between soil particles must be saturated or near saturated.

lithosphere: the crust and upper mantle of the Earth

livelihood: job or skill that supports a person's existence, so that they can have the necessities of life

low-density housing: residential developments with around 12–15 dwellings per hectare, usually located in outer suburbs

mantle: the layer of the Earth between the crust and the core

medium-density housing: residential developments with around 20–50 dwellings per hectare

megacity: city with more than 10 million inhabitants

megaregion: area where two or more megacities become connected as increasing numbers of towns and ghettos develop between them

metropolitan region: an urban area that consists of the inner urban zone and the surrounding built-up area and outer commuter zones of a city

microclimate: specific atmospheric conditions within a small area

migrant: a person who leaves their own country to go and live in another

migration: the movement of people (or animals) from one location to another

moraine: rocks of all shapes and sizes carried by a glacier

nomadic: describes a group that moves from place to place depending on the food supply, or pastures for animals

Pangaea: the name given to all the landmass of the Earth before it split into Laurasia and Gondwana

peninsula: land jutting out into the sea

per capita income: average income per person; calculated as a country's total income (earned by all people) divided by the number of people in the country

perennial: describes a stream that flows all year

permafrost: a layer beneath the surface of the soil where the ground is permanently frozen

physical process: continuing and naturally occurring actions such as wind and rain

plateau: an extensive area of flat land that is higher than the land around it. Plateaus are sometimes referred to as tablelands.

population density: the number of people living within one square kilometre of land; it identifies the intensity of land use or how crowded a place is

population distribution: the pattern of where people live. Population distribution is not even — cities have high population densities and remote places such as deserts usually have low population densities.

prevailing wind: the main direction from which the wind blows

pull factor: favourable quality or attribute that attracts people to a particular location

push factor: unfavourable quality or attribute of a person's current location that drives them to move elsewhere

quality of life: your personal satisfaction (or dissatisfaction) with the conditions under which you live

rain shadow: the drier side of a mountain range, cut off from rain-bearing winds

rebates: a partial refund on something that has been bought or paid for

resource: something that is of use to people

retrofitting: adding a component or accessory to something that did not have it when it was originally built or manufactured

rift zone: a large area of the Earth in which plates of the Earth's crust are moving away from each other, forming an extensive system of fractures and faults

rip current: a strong localised current that channels a fast-flowing stream of water offshore

sanitation: facilities provided to remove waste such as sewage and household or business rubbish

sastrugi: parallel wave-like ridges caused by winds on the surface of hard snow, especially in polar regions

sea change: movement of people from major cities to live near the coast to achieve a change of lifestyle

sediment: material carried by water

seismic waves: waves of energy that travel through the Earth as a result of an earthquake, explosion or volcanic eruption

selective logging: a forestry practice in which only selected trees are cut down

shifting agriculture: process of moving gardens or crops every couple of years because the soils are too poor to support repeated sowing

slums: a run-down area of a city characterised by poor housing and poverty

soluble: able to be dissolved in water

species: a biological group of individuals having the same common characteristics and able to breed with each other

stalactite: a feature made of minerals, which forms from the ceiling of limestone caves, like an icicle. They are formed when water containing dissolved limestone drips from the roof of a cave, leaving a small amount of calcium carbonate behind.

stalagmite: a feature made of minerals found on the floor of limestone caves. They are formed when water containing dissolved limestone deposits on the cave floor and builds up.

subsistence: producing only enough crops and raising only enough animals to feed yourself and your family or community

sustainable development: economic development that causes a minimum of environmental damage, thereby protecting the interest of future generations

swash: the movement of water in a wave as it breaks onto a beach

tectonic plate: one of the slow-moving plates that make up the Earth's crust. Volcanoes and earthquakes often occur at the edges of plates.

temperate: describes the relatively mild climate experienced in the zones between the tropics and the polar circles

transportation: the movement of eroded materials to a new location by elements such as wind and water

treaty: a formal agreement between two or more countries

tree change: movement of people from major cities to live near the forest to achieve a change of lifestyle

tributary: river or stream that flows into a larger river or body of water

triple bottom line: an accounting term for measuring the success of a city, country or organisation by the health of its environment, its society and its economy

undertow: the powerful flow of water in the waves as it returns to the sea

urban: relating to a city or town. The definition of an urban area varies from one country to another depending on population size and density.

urban sprawl: the spreading of urban areas into surrounding rural areas to accommodate an expanding population

urbanisation: the growth and expansion of urban areas

utilities: services provided to a population, such as water, natural gas, electricity and communication facilities

viable: capable of working successfully

volcanic loam: a volcanic soil composed mostly of basalt, which has developed a crumbly mixture

weathering: the breaking down of bare rock (mainly by water freezing and cooling as a result of temperature change) and the effects of climate

INDEX

ACROS Fukuoka building   230, 249
solar power   226

K

Kakadu National Park   36, 37–8
karst landscapes   19, 31–2
katabatic winds   97
Kati Thanda–Lake Eyre   23, 24–5, 73
Kuku Yalanji people   145

L

Lake Mungo   93–5
lakes   24–5
land features, recognising   27
landforms
    in the Pacific   27
    regions in Australia   22–3
landscape protection   39–41
landscapes
    different cultural perspectives
        on   36–8
    formed by ice   71
    formed by water   44–6
    processes shaping   27–9
    types   18–20
landslides   157
Late Cretaceous period   144
lateral plate slippage   112
latitude   81
leaching   142
limestone stack   45, 54
lithosphere   113, 140
Little Sandy Desert   91
Loading Dock, Baltimore   227–8
logging   152
longitude   81
longshore drift   50, 51, 56
low-density housing   230
lowland tropical rainforests   139

M

mantle (Earth)   28
Mariana Trench   111–2
meanders   67, 68
medium-density housing   230
megacities   214–16
megaregions   215
microclimate   141
Mid-Atlantic Ridge   126
migration *see* internal migration;
        international migration to Australia
mining
    in Antarctica   99–100
    deforestation and   153
Mirrar people   38
Mississippi drainage basin   69
Mississippi River   69–70

Mississippi River and Tributaries
        Project   70
Modified Mercalli scale   119
Mount Taranaki   128–30
mountains   19
    climate and weather   106
    dome mountains   115
    in Dreaming Stories   108–9
    fault-block mountains   114–15
    fold mountains   113–14
    formation   111–17
    global distribution   104–5
    highest   104
    plateau mountains   117
    population density   107
    sacred and special places   109
    survival skills   109–10
    types   113–17
    use by people   107–10
    volcanic mountains   125–7
Murray River
    mouth   51
    tributaries   67
Murray–Darling Basin   22, 24, 51

N

Nambiquara people, Brazil   156
national parks   159
Nepal, 2015 earthquake   119–20
New York City   222–3
North American Plate   112, 126
Nullarbor Plain   31

O

ocean trenches   112
ocean waves
    and coastal erosion   47–8
    constructive waves   49
    destructive waves   47
Okavango River   71
orangutans, deforestation and   155
Ota, Japan   226

P

Pacific Ring of Fire   112
Pangaea   116
Pearl River delta   215
Pedirka Desert   91
per capita income   210
perennial streams   64
permafrost   19
Perth, WA   11, 12
photographs, describing   202
physical processes, definition   56
pictographs   184
place, as geographical concept   5
plateau mountains   117
plateaus   19

playas   88
polar deserts   84
polar regions   19
pollution
    air pollution   182–3, 211
    transport and   211
population density
    Australia   168–70
    average for each continent   170
    definition   168
    disease and   182
population distribution
    Australia   6, 168, 172, 177
    definition   168
population growth, Australia   189
population profiles   185
positional language   35
prevailing winds   50
public transport
    benefits   247
    investment in   243
pull factors, rural–urban
        migration   194, 203
push factors, rural–urban
        migration   194–5, 203

Q

quality of life   235

R

rain shadows   82
rain-shadow deserts   82–3
rainfall, annual average in
        Australia   6
rainforests
    Amazon rainforest   146–9
    in Australia   144–6
    benefits of   149–50
    causes of deforestation   152–3
    conservation   159–60
    definition   19, 138
    deforestation   152–3
    ecosystems   141–2
    impacts of deforestation   154–8
    indigenous people and   144,
        149–50
    location   138
    physical processes   140
    tropical rainforests   138–40
    types   139
Ranger uranium mine   38
Regional Sponsored Migration
        Scheme (RSMS)   190
resources   52
Richter scale   119
rift zones   126
rip currents   62
river landscapes, formation   64–8
river management   69–70